Practical Radiobiology for Proton Therapy Planning

Series in Physics and Engineering in Medicine and Biology

About the Series

Series in Physics and Engineering in Medicine and Biology will allow IPEM to enhance its mission to 'advance physics and engineering applied to medicine and biology for the public good.'

Focusing on key areas including, but not limited to:

- clinical engineering
- diagnostic radiology
- informatics and computing
- magnetic resonance imaging
- nuclear medicine
- physiological measurement
- radiation protection
- radiotherapy
- rehabilitation engineering
- ultrasound and non-ionising radiation.

Practical Radiobiology for Proton Therapy Planning

Bleddyn Jones
Gray Laboratory, CRUK/MRC Oxford Oncology Institute, The University of Oxford, ORCRB-Roosevelt Drive, Oxford OX3 7DQ, United Kingdom

IOP Publishing, Bristol, UK

ISBN 978-0-7503-1338-4 (ebook)
ISBN 978-0-7503-1339-1 (print)
ISBN 978-0-7503-1340-7 (mobi)

DOI 10.1088/978-0-7503-1338-4

Version: 20171201

IOP Expanding Physics
ISSN 2053-2563 (online)
ISSN 2054-7315 (print)

British Library Cataloguing-in-Publication Data: A catalogue record for this book is available from the British Library.

Published by IOP Publishing, wholly owned by The Institute of Physics, London

IOP Publishing, Temple Circus, Temple Way, Bristol, BS1 6HG, UK

US Office: IOP Publishing, Inc., 190 North Independence Mall West, Suite 601, Philadelphia, PA 19106, USA

This book is dedicated to Dr O C A Scott, MD (Cantab) FRCR, who inspired so many scientists and clinicians, gave freely of his ideas and was such a generous benefactor.

Contents

Preface

This book is intended to cover the principles concerning the advantages and potential pitfalls that occur in proton therapy, especially its radiobiological modelling applications. For these to be understood, it is essential to present summaries of radiotherapy, radiobiology, clinical radiobiological modelling for conventional radiotherapy techniques as well as for protons and other ion beams. Radiobiological modelling has proved useful in radiotherapy, especially in situations where departures from protocols occur for a variety of reasons such as unintended treatment interruptions, errors in treatment delivery, dose rate effect applications, comparisons of different techniques, dose fractionation schedules, re-treatments and even in clinical trial design. These techniques, if suitably adapted, are capable of producing similar insights and practical guidelines in proton and other forms of ion beam therapy. The optimisation of proton and ion beams is essential since there is considerable 'competition' from the more advanced forms of megavoltage photon-based radiotherapy, which allows fewer and highly focused treatments to be given as well as being more economical.

The use of relatively simple mathematics and some very basic worked examples are included within the text, but have been kept to a minimum, so that *all the relevant disciplines* can understand the principles, and then use the historical content and advice provided to develop the subject further. A basic knowledge of radiation biology is assumed, and references have been kept to the minimum necessary. The parameters chosen for exploratory modelling and illustrative purposes may require modification and further input for more specific clinical applications. Any errors are the sole responsibility of the author. There is inevitably some degree of repetition where this is considered essential.

The mainstay approach has been to use the biological effective dose (BED) concept based on the linear-quadratic (LQ) model of radiation effect. In recent years it has been possible to assess the BED values of charged particle therapies such as protons and light ions by incorporation of the maximum and minimum limits of the relative biological effect (RBE), which accounts for the increasing complexity of DNA damage and the increasing proportion of non-repairable damage that occurs when the linear energy transfer (LET) of a radiation is increased. Typically LET increases within the Bragg peak region for charged particles, with resultant increments in the RBE, but the RBE reduces with increasing dose and also at very high LET values, but it is important to realise that different biosystems will have quite different RBE values. The topic of the RBE remains controversial, although it is being increasingly understood, and there are several models that attempt to link LET with the RBE, many of which use formidable mathematics. In contrast, this book addresses how to achieve this by using relatively simple mathematics. Such a simpler approach can be understood by physicians, physicists and biologists. More complex approaches can be limited in their usefulness due to excessive and/or restrictive assumptions and the limits of what can be known. Simpler models,

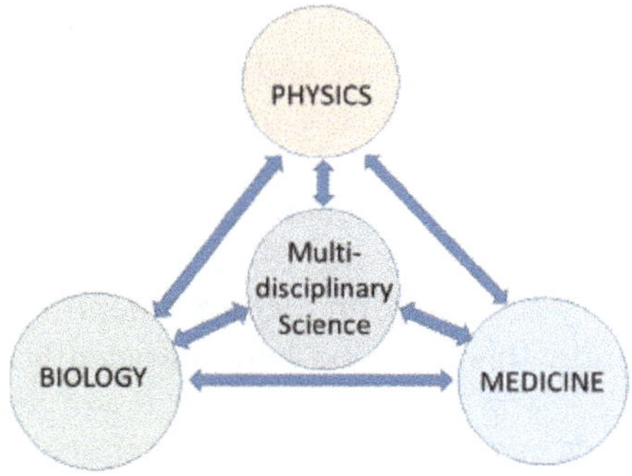

Figure 1. The central multi-disciplinary role, which must include a modelling component.

although they always require caveats, are capable of providing useful qualitative and quantitative understanding of the basic principles that govern clinical outcomes.

This subject is very multidisciplinary, and is based on a triangular arrangement with physics, medicine and biology forming the apices of an isosceles triangle (figure 1). The decisive middle ground has to be covered by individuals who have sufficient knowledge and experience of all three disciplines. A basic knowledge of these three subjects and the inter-disciplinary subject of radiation biology is assumed, along with basic aspects of particle therapy such as the Bragg peak effect, and the characteristics of the spread out Bragg peak and how this may be achieved. The author is minded that the text should be reasonably understood by members of all these disciplines, although additional reading will be required from review articles and textbooks to supplement cross-disciplinary understanding. Also, each primary discipline is divided into a spectrum of further specialisation groups: physicists are split into fundamental particle physicists, accelerator or detector physicists and their medical physicist colleagues; in biology there are purely molecular biologists (biochemists) concerned with *in vitro* work, biophysicists (physiologists), pharmacologists and zoologists, some concerned with *in vivo* animal testing; within the faculties of medicine, the knowledge base is spread over anatomy, all forms of pathophysiology, oncology, surgery, patient management as well as a broad interest in epidemiology. Also, many non-biologically trained physicists have had a very significant influence in advancing radiobiology and, given the added complexities of charged particle physics, their continued involvement will be essential to understanding particle therapy and its associated radiobiology.

With all these facets in mind, it is necessary to write sympathetically for the non-specialist, but furnish sufficient detail regarding many necessary fundamental aspects across these topics, while providing more detailed information for the practitioner or researcher in this field. For reasons of space and economy, it has been necessary to omit the classical graphs and diagrams associated with the basic sciences, but these can be found in standard textbooks. In most instances, since most readers of this book will be familiar with them, the brief qualitative descriptions and definitions provided will act as an aide memoire.

The medical decision makers of the future in particle therapy will need broad scientific backgrounds, excellent clinical and oncological training, and be able to appreciate the strengths and weaknesses of the available mathematical models that link the clinical, biological and physical parameters, and which can be used to protect the patient against over or under-dosage. Modelling is a necessary way of quantitative communication between the three apices already referred to, in the important multidisciplinary centre-ground, complementing rational verbal or written approaches. In essence, this must be kept sufficiently simple in order for all disciplines to interact.

Considerable practice is necessary to become familiar and competent in this subject. The calculations may appear to be deceptively easy, as only knowledge of some higher school mathematics is necessary, and this can be aided by computer software systems such as Mathematica, Matlab and Maple. However, some of the clinically based calculations are quite protracted, with many more pitfalls (due to the RBE, LET, etc) than for similar calculations in the case of photon-based treatments.

There are many who regard particle therapy as just a seamless extension of photon-based radiotherapy. This is certainly useful in terms of departmental organisation within a hospital structure, to ensure good patient access and to capitalise on site-specialist knowledge of the oncologist in terms of detailed regional anatomy and indications for therapy. However, owing to the added complexity of particle therapy, there is an essential need to educate and train all the sub-disciplines involved in treatment delivery and patient management in order to inform them of the strengths and weaknesses of particle therapy, and by knowing these to continually search for better, highly optimised therapy.

Depending on the background of the reader, it is not necessary to read the whole of the book. Chapter 1 provides the essential physics background mostly for the benefit of biologists and clinicians, but includes the importance of the choice of the reference radiation in RBE studies as well as the different ways in which LET is expressed, and introduces the potentially important parameter of inter-track distance. In chapter 2, the basis of radiobiological modelling is introduced, with formulations for conventional megavoltage photon radiotherapy and for particle therapy, including low and high dose, dose rate variations and the useful BED concept. Chapters 3 and 4 have been written for persons who have little or no previous experience of radiotherapy, with discussion of the essential medical aspects of treatment planning, including the influence of surgery and other factors on tissue viability. Some of the key historical developments in radiotherapy are covered in chapter 5, as well as what has been learned from extensive basic and clinical research using fast neutrons for modern applications with charged particles. Chapter 6 considers dose-fractionation effects in photon and particle therapy, with some worked examples. The rationale for using a variable as opposed to the conventional constant RBE in proton therapy is provided in chapter 7. This is followed by the description of a relatively simple 'energy efficiency' model for estimation of the RBE values for any ion beam in chapter 8. More specific applications in proton therapy are given in chapter 9, with some tables that suggested tentative RBE allocations for variable values of LET in different classes of tissues and tumours. The correction of unintended treatment interruptions for high LET treatments, with worked examples, is described in chapter 10, which also includes formulations for assessing re-treatment doses in the central nervous system. Chapter 11 considers the radiobiological correction of Bragg peak placement errors, resulting in deviations in both LET and dose. Chapter 12 contains further possible future developments to improve our overall understanding and especially to obtain more accurate modelling parameters, as well as methods of assessing the potential impact of concomitant drug sensitisation with high LET radiations.

It is hoped that this book will not only improve the safety and effectiveness of particle therapies, but also inspire further research and enquiry.

Acknowledgments

I am extremely grateful for the help and encouragement provided by the following.

1. Many colleagues who have inspired me over many years to seek solutions to high LET problems, too large a number to mention individually, but especially Roger Dale, Gillies McKenna, John Hopewell, Oliver Scott, Jack Fowler, Mark Hill, Boris Vojnovic, Ken Peach, Claire Timlin, Herman Suit, Peter O'Neil, Dudley Goodhead, Kevin Prise, Stuart Green, Karen Kirkby, David Colling, Douglas Errington, Andrzej Kacperek, Richard Britten and Hilmar Warenius for stimulating discussions and encouragement.
2. CERN, Geneva for a Visiting Scientist award during 2014, and to the Director General for the award of Guest Professor 2015–2016, with the support of Professor Manjit Dosanjh.
3. The Principals and Fellows of Brasenose College, Oxford during 2010–2015, especially Sir Roger Cashmore, and Green Templeton College as from 2017.

The author has been an investigator on several UK Research Council and EU FP-7 funded grants concerned with particle therapy, including ENVISION (241851), ENTERVISION (264552), and ULICE (228436), and would also like to express thanks to The Medical Research Council UK for support to study radiobiology (1980–1983), and to The Cyclotron Trust (2000–2005).

I'm also grateful to all the journals that have published my work on this subject, and for reproduction of graphics, some of which are included in this book.

Glossary of terms and symbols

α, the radiosensitivity coefficient of the LQ model per unit dose, in units of Gy^{-1}.

α_L, refers to α for the low LET condition when high and low LET qualities are being compared.

α_H, refers to α for the high LET condition when high and low LET qualities are being compared.

α_U, the ultimate value of α at the turnover point of the LET–RBE relationship.

Similarly, β is the radiosensitivity coefficient of the LQ model per unit dose squared, in units of Gy^{-2}.

β_L, β_H, and β_U refer to the same specifications as for α.

α/β, this important ratio, which has units of Gy, determines the dose per fraction sensitivity in an inverse sense, such that low values (2–3 Gy) show large dose per fraction sensitivity, but high values denote low dose per fraction sensitivity (those above 7 Gy).

$(\alpha/\beta)_L$, refers to the low LET ratio in situations where low and high LET radiation qualities are being compared.

$(\alpha/\beta)_H$, refers to the high LET ratio in situations where low and high LET radiation qualities are being compared.

RBE, defined as the ratio of dose for low LET (the control or reference) radiation divided by the dose of the high LET test radiation in order to achieve the same specified bio-effect.

RBE_{max}, the limiting value of RBE at very low dose (approaching zero dose) for any specified high LET radiation when compared with a low LET radiation.

RBE_{min}, the asymptotic limiting value of RBE at very high dose for any specified high LET radiation when compared with a low LET radiation.

RBE_C, an intermediate value of the RBE where the following conditions apply: $R_{max}/R_{min}^2 = (\alpha/\beta)_H/(\alpha/\beta)_L$.

LET, the linear energy transfer parameter defined as the energy release per unit length of track in the track plane expressed in units of keV μm^{-1}. Further definitions are given in chapter 1.

LET_U, the value of LET at the turnover point of the LET–RBE relationship.

LET_C, the value of LET of the low LET control radiation quality.

LET_x, any specified value of LET for the test high LET radiation.

N, the number of dose fractions in a treatment course (or schedule).

d, the dose per fraction in units of Gy, D is the total dose (or $n \times d$).

E, the negative logarithm of the surviving fraction, often referred to as the log cell kill, which is proportional to the surviving fraction.

SF, the surviving fraction of cells in an experiment.

g, the ratio of a specified normal tissue dose divided by the tumour dose.

S, a formulation for combining the g factor with the uncertainties of knowledge regarding the most appropriate RBE value in normal tissues and tumour.

BED, the biological effective dose, in units of Gy, which expresses the dose required to achieve a specific bio-effect if given in ultra-small dose fractions, and represents a ceiling of dose that can be used for comparative purposes.

ω, the average cellular doubling time during radiation.

K, the BED equivalent of cellular repopulation expressed as Gy day^{-1}.

T_K, the lag time, in days, that may occur before cellular repopulation is manifest.

f, refers to the mean inter-fraction interval (days). For example, if five treatments per week are given $f = 7$ days/$(5 - 1) = 7/4 = 1.75$ days.

Z, the nuclear charge.

z, the tumour dose if the normal tissue dose is specified as d.

μ_F, the DNA repair rate constant for the fast component of repair.

μ_S, the DNA repair rate constant for the slow component of repair.

h, the incomplete repair parameter for closely spaced treatments.

Fluence refers to the number of radiation tracks crossing a specified area.

KERMA refers to the kinetic energy released per unit mass (J Kg^{-1}).

Absorbed dose refers to the energy absorbed in a specified mass (J Kg^{-1}).

Radiation quality refers to the LET of a radiation at a specified energy.

Author biography

Bleddyn Jones

Bleddyn Jones studied Medicine at Cambridge (with some mathematics), followed by clinical and post-graduate training at Guy's, St Thomas, St Bartholomew's and The Royal London Hospitals. He also studied Radiobiology (University of London) before commencing Clinical Oncology training, subsequently holding Consultant and Academic appointments in Clatterbridge, Hammersmith, and Birmingham. He later became Professor of Clinical Radiobiology in Oxford, where his research concentrated on developing mathematical models of particle therapy, as well as other difficult aspects of radiotherapy such as re-treatments and radiosurgery. This was an extension of his previous applied radiobiology work in reducing toxicity following Gynaecological Brachytherapy, initially using models developed by Professor RG Dale, with whom he collaborated extensively over many years.

IOP Publishing

Practical Radiobiology for Proton Therapy Planning

Bleddyn Jones

Chapter 1

Particle physics for biological interactions

1.1 Physical beam parameters, essential dosimetry and reference (or control) radiation requirements for RBE studies

This chapter, intended more for the benefit of biologists rather than physicists, includes a short qualitative description of the physics base of photon and particle therapy, including linear energy transfer (LET) and its variants, and finally the choice of the control or reference radiation for relative biological effect (RBE) studies. It is assumed that the reader will be familiar with Bragg peaks and how these peaks are spread out to cover a defined clinical target volume. A comprehensive account of the physics of proton therapy by Newhauser and Zhang (2015) is appropriate only for physicists, but their figure 15 on page 179 is reproduced here as figure 1.1, and shows the dose (D) for multiple Bragg peaks obtained by using different proton energies, resulting in peaks at different depths (indicated by z on the abscissa) to form a plateau of relatively uniform dose when they are all summated. It should be noted that the flat section of the green line is composed of low LET radiation with the high LET being within the Bragg peak and the maximum LET at its end. The reader is encouraged to refer to this diagram while reading this chapter.

First, it is necessary to understand the basic concepts that underpin radiation dosimetry and the important differences between photon and positive charged particle therapies. For those less familiar with radiotherapy physics, or who wish to revise the key aspects, a brief summary is provided, concentrating on important definitions that lead up to the concept of LET, and its various forms. For more complete descriptions the reader should consult standard textbooks.

Radiation fluence refers to the number of radiation beamlets (or individual tracks) (N) within a beam which crosses the unit surface area, that is N/A, where A is the area covered by the whole beam, and so the units are cm^{-2}.

KERMA signifies the kinetic energy released per unit mass of a specified mass of medium, by release of secondary particles such as δ-rays and all ionisation products.

doi:10.1088/978-0-7503-1338-4ch1

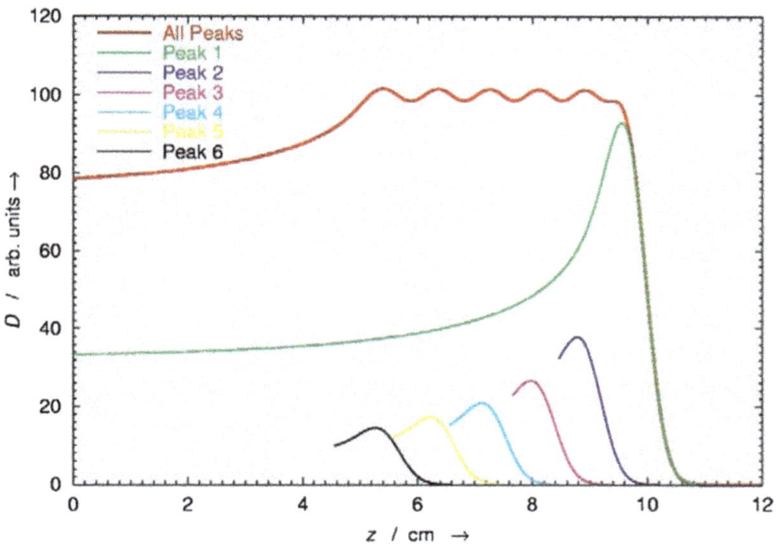

Figure 1.1. Absorbed dose D is shown as a function of depth z in water from a spread-out Bragg peak (SOBP) (uppermost curve) and its constituent pristine Bragg peaks (for the lower curves, for clarity; all but the deepest pristine Bragg peak are only partly drawn). In many cases, the clinical target volume is larger than the width of a pristine Bragg peak. By appropriately modulating the proton range by using different energies from the highest (peak 1), which provides dose to the greatest depth, and then with progressively reduced energies to provide peaks at each of the lower depths, the lowest energy providing peak 6. The fluence (here proportional to the height of each peak) is also varied in order to provide the intended uniform dose to the target, although in this example there is some undulation of the 'all peaks' summated dose from each separate energy contribution because of the low number of peaks used for illustrative purposes. Reproduced from Newhauser and Zhang (2015).

However, some of the energy released (in J Kg^{-1}) will be sufficient to pass beyond the volume of the mass of interest and be absorbed elsewhere. Consequently it was necessary for the physicist L H Gray to define the **absorbed dose** as the energy absorbed by a mass of interest, again as J Kg^{-1}, which should include not only the primary ionisation events, but also all secondary particles, such as δ-rays, neutrons, recoiled neutrons/protons and fragmented ions, that make a contribution.

Most dose calculations are performed for unit density water, so mass and volume have the same number. For denser structures, such as bone or lower density air cavities, the absorbed dose is larger and smaller, respectively, and is often assumed to be mainly dependent on electron density in the case of megavoltage x-rays (photons) in the most commonly used clinical range (4–8 MeV). Lower energy x-rays (especially below 100 keV, which may form part of the x-ray spectrum from x-ray generators that have higher peak energies such as 200–300 keV) cause ionisation by the photoelectric effect in proportion to the atomic charge cubed (i.e. Z^3). In contrast, uncharged fast neutrons preferentially interact with neutron deficient nuclei such as hydrogen, producing recoil protons, so that richly hydrogenated compounds, such as lipid rich tissues (such as myelin in brain white matter and fat) have a greater KERMA than water.

To arrive at the absorbed dose, use is made of energy transfer coefficients and radiation absorption parameters. It is important to note that the loss of dose with depth for divergent radiation, for example x-rays emanating from a high-Z metal target, are mostly due to inverse square law (ISL) effects as well as due to tissue absorption. For highly collimated beams, such as pencil beams, ISL effects are much reduced.

Particle beams may be delivered as pencil beams, but may also be passively scattered with resultant divergence and ISL effects. There is a further divergence effect in all cases, caused by coulombic force scattering of similarly positively charged particles (when stripped of their electrons), which also depends on their energy and mass (each of which reduces the scattering when large). The resultant diversion of the beam with tissue depth reduces the fluence.

In particle therapy the intended absorbed dose is normally prescribed as an equivalent dose. This can cause confusion to practitioners, since the equivalent dose has been expressed in different ways, for example the cobalt equivalent Gy, or Gy equivalent (or eq-Gy), or Gy-RBE or RBE-Gy, where the particle dose has been reduced by the working RBE value in order to achieve the same intended effect as from, say, megavoltage x-rays or photons. Further confusion is caused by the standard expression for equivalent Gy, for example, as

$$\text{Gy–RBE absorbed dose} = \text{particle absorbed dose} \times \text{RBE},$$

but it is important to realise and understand the re-arrangement, which is used in practice, as

$$\text{particle absorbed dose} = \text{Gy–RBE absorbed dose} \div \text{RBE}.$$

The physician chooses the most appropriate eq-Gy (or RBE Gy) dose of photons (or x-rays); this is then divided by the RBE to provide the physical absorbed dose of particles given to the patient.

Very recently, the ICRU has decided not to continue the use of the RBE-Gy, which was especially associated with the constant 1.1 RBE used for proton therapy. Doses should now be specified in Gy for each radiation modality, which will allow better identification of RBE values in the future.

In routine clinical megavoltage x-ray (photon) therapy, the ionisation is relatively uniform along the radiation track. The depth dose curve for a monoenergetic and non-divergent beam shows a fairly constant fractional loss of absorbed dose per unit distance traversed from its initial maximum ($d_{\max}$) dose so that at deeper distances (denoted by x) the dose d_x can be approximated as $d_x = d_{\max}e^{-\mu x}$, where μ is the attenuation coefficient. The same is not true along a charged particle beam since slowing of the particle increases the energy loss (E) per unit distance (x), denoted by $\mathrm{d}E/\mathrm{d}x$. This is because of the increasing probability that opposite coulombic charges will interact, causing more ionisation and so increased KERMA and locally absorbed dose, the extreme case being at the distal part of the Bragg peak itself. Since the ionisation probability changes with depth it is useful to consider a further parameter, the LET, which reflects the closeness of local ionisations along a beam track. LET is not only relevant to charged particles but also to low energy photons/x-rays since whenever photon energy falls to the keV range their secondary particles have lower energy and a higher LET.

LET is defined as the energy lost per unit length of medium by a charged particle (as, e.g., 1 keV μm^{-1}, or 1.602 J m^{-1}). There are several variants of LET such as the following.

1. L_Δ, where Δ refers to the maximum limit of energy (e.g. L_{100} would consider only energies below 100 keV μm^{-1}) and LET as total energy loss (L_∞). This latter parameter reflects 'stopping power' in the medium, and so includes its density, with units expressed as MeV cm^2 g^{-1}, or J m^2 kg^{-1}.
2. If there are different energies in a beam, as might be the case with fast neutrons or SOBP charged particles, a LET spectrum can be used, calculated as either 'track average' or as 'absorbed dose average' (or energy average) LET.
3. In microdosimetry, the unsatisfactory aspects of 'average LET' are often overcome by graphical presentations of LET plotted against the dose fraction per log LET interval.
4. The lineal energy (y), which takes account stochastic energy deposition (whereas LET does not). This is defined as $y = \varepsilon/d_{av}$, where ε is the energy imparted in a volume with d_{av} being the mean track chord length in the volume.
5. The final variant, with attractive properties for bio-molecular considerations, is to account for δ-rays ejected from tracks. These δ-rays are radially distributed and responsible for most bio-effects and ionisations collected by detectors. With considerable acumen, Katz *et al* (1971) proposed the use of $Z^{*2}\beta^{-2}$, where Z^* is the effective nuclear charge of the atomic nucleus of atomic charge Z and β is the relativistic velocity (v/c), in order to estimate the effective radius of ionisation, which should correlate with bio-effectiveness. It must be realised that, as fully stripped ions slow down, they pick up electrons while causing ionisation, so Z^* becomes less than Z, the parameter will change with depth. This is discussed further in chapter 8.

The loss of energy E with distance in a medium is well described by the Bethe Bloch equation:

$$-\frac{\mathrm{d}E}{\mathrm{d}x} = \frac{4\pi}{m_e c^2}\frac{nz^2}{\beta^2}\left(\frac{e^2}{4\pi\epsilon_0}\right)^2\left[\ln\left(\frac{2m_e c^2\beta^2}{I(1-\beta^2)}\right) - \beta^2\right].$$

For the purpose of understanding, this formidable formulation can be reduced to:

$$-\mathrm{d}E/\mathrm{d}x \text{ (energy/cm)} = K \text{ (charge}^2\text{/velocity}^2\text{)},$$

where K replaces the constant physical parameters. It can easily be seen that, for a proton (charge = 1), or carbon (charge = 6, if all electrons are stripped away), the energy loss will depend on the Z number but will increase sharply as velocity falls. This occurs as particles slow down through a medium. After such slowing they interact with orbital electrons in nearby molecules and cause ionisation. The Bragg peak represents a sharp rise in dE/dx with loss of velocity or energy at increasing depth. The initial energy or velocity will govern the complete range (or depth) of energy deposition, so beams with higher incident energies at the level of the skin will have Bragg peaks at greater depths.

1.1.1 Straggling and fragmentation

This term is used to describe the variation in path length caused by differences in energy and velocity reduction through a medium or tissue, due largely to differences in interactions along the beam path due to variations in electron density/tissue composition and other stochastic events. It normally denotes a blurring-like effect on the beam towards its distal edge, but there is also some degree of lateral straggling for particles that are more easily scattered. Thus, the beam edge is not so sharp with protons, which are lighter than helium or carbon ions. For even heavier ions, such as carbon, there is a distinct fragmentation tail caused by nuclear interactions and emitted γ-rays, which results in a tail of dose beyond the end of the Bragg peaks. For lighter particles, such as protons, fragmented ions do not occur but, as with heavier particles, the tissue interactions can cause emission of γ-rays and some neutrons.

1.1.2 Separation of charged particles with increasing tissue depth

The coulombic forces due to the nuclear charge will cause charged particles to separate from each other gradually in proportion to their tissue depth, and also in inverse proportion to their mass and velocity. This will occur in a beam traversing a vacuum. This is often referred to as 'coulombic scattering', but in matter will also include much wider angle scattering due to repulsive charge interactions with atomic nuclei (as in the classical example of Rutherford scattering). So, protons (charge = 1, mass unit = 1) will separate less than electrons (charge = 1, mass $\approx$ 1/2000 of mass unit), Compared to protons, heavier ions (also containing neutrons so they have mass greater than their Z number) show lesser deviation despite their increased charge, even if fully 'stripped' of electrons before their acceleration, although some particles will gain electrons during their transit through matter.

There appears to be no formal term that describes the average or median inter-track separation distances, which may contribute to biological effect changes over and above the LET, as discussed below. It is easy to crudely estimate the mean inter-track separation distance (s) by the following basic approach (see figure 1.2).

The overall area A is divided by the number of tracks to provide A/N, the average area surrounding each track; if this is assumed to be a small square area (represented by s^2), then $A/N = s^2$. If, on average, the track occupies the centre of each square, then the average track separation will be s, which is $\sqrt{(A/N)}$, and which can also be expressed as $\sqrt{\frac{1}{F}}$, where F is the fluence (defined above). This might be an important parameter to use in the analysis of cell-survival studies and treatment outcomes, as it could modify the effect of LET and dose. For example, even if the LET parameter is calculated to be the same at, say, the mid-Bragg peak for the same particle type but accelerated to two quite different energies resulting in Bragg peaks at two different depths (where the particle energies will then be the same), any reduction in bio-effectiveness could be due to greater separation of tracks in the case of the particles

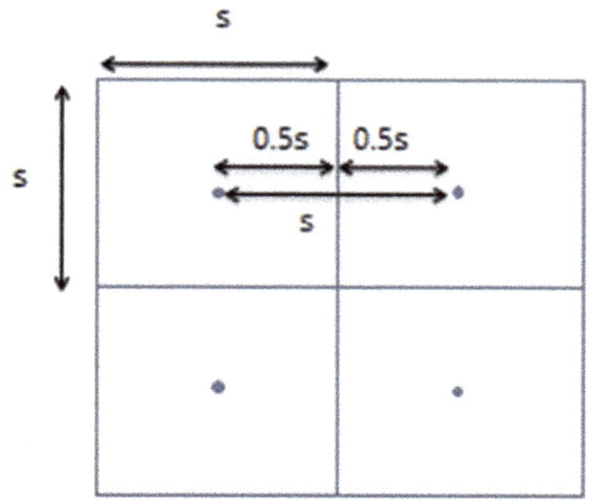

Figure 1.2. Four particle tracks are represented in cross section by central points within a square of side s; the inter-track distance is then s. For N tracks, the overall area (A) is Ns^2.

that were accelerated to the higher energy. Such a process would depend on increasing track separation due to coulombic scattering, and even more so on the deliberate use of scattering filters, causing ISL geometric dispersion of the beam at increasing depths.

This can be further expressed in a very simple way. Consider tracks of radius a, with volume $\pi a^2 x$ when considered as a cylinder extending to depth x. The inter-track distance is s, so that the fractional volume occupied by a single track compared with its surrounding volume of $s^2 x$ is

$$\frac{\pi a^2 x}{s^2 x} = \frac{\pi a^2 N}{A} = \pi a^2 F$$

(since $s^2 = 1/F = A/N$).

For the same fluence, as the overall area (A) of a beam reduces, the closeness of tracks per unit area (and volume) increases. But if A increases, the fluence falls and the closeness of tracks decreases. For different ions the specific Katz track radius may be used.

Such an effect can actually be visualised using GEANT4 and FLUKA in-silico simulations, shown to the author by Dr David Colling (Imperial College) and Dr Francesca Fiorini (Oxford), respectively. Although LET may be similar for each track at each depth, their increased separation may be sufficient to diminish the bio-effect since more cell nuclear material or even whole cells may be un-irradiated. Some evidence for reduced bio-effectiveness with depth in the case of passively scattered beams is given below.

1.1.3 Particle accelerators

Synchrotrons and cyclotrons are the commonest accelerators used for charged particle medical beams. These terms can confuse since modern cyclotrons are in fact synchro-cyclotrons, but the more basic nomenclature remains useful. Printing space does not permit a full description of these. For a summary of how these accelerators operate the reader should refer to short accounts such as Peach *et al* (2011), designed for hospital readers, with more complete accounts by Wilson (2001) in his excellent textbook, which contains ample worked examples concerning beam control, etc. Suffice it to say here that cyclotrons operating from a circular cavity arrangement accelerate protons to their peak energy on extraction. To produce lower energies that result in Bragg peaks at depths less than the maximum range, the beam must be degraded within metal or plastic strips of differing thicknesses (which become highly radioactive and must be situated sufficiently remote from the patient because of induced radiation including neutrons). Cyclotrons deliver high dose rates (0.5–1 Gy s^{-1}). For example 12 Gy in 24 s is not unusual.

In contrast, synchrotrons, which are used to accelerate protons or particles heavier than helium, occupy a much larger floor space and operate by a single radiofrequency accelerator combined with magnetic focusing and defocusing bending the beam around a geometrical arrangement of magnets, which leads the beam back to the original point for further acceleration. The magnetic field strength must be carefully increased to match the particle energies at each 'cycle'. When the desired energy is obtained the 'particle bunch' is extracted and then delivered.

There are two options for controlling beam size: the pencil beams delivered from the accelerator can be passively scattered by a filter to a wider area, but with a small neutron contamination, depending on the efficiency and composition of the filter. The alternative is to use the pencil beam and magnetically deflect the beam to the positions required (active scanning), which avoids neutron production and provides better conformity of dose to the intended targets.

In each case the energies delivered from these accelerators have a spread in terms of energy and in spatial properties. There are some useful terms to denote these.

Luminosity refers to the total energy emitted, often expressed per unit time.

Emittance measures the average spread of particle coordinates with respect to their position in 3D space and their momentum, with dimensions of length (e.g. metres) or length times angle (metres times radians). In a low-emittance particle beam, the particles are confined to a small distance and have nearly the same momentum. The probability of particle interactions will be greater, resulting in higher luminosity.

Also, further beam parameters include energy spread (in standard deviations), pulse repetition rate (f), bunches per macropulse (Mb) and effective bunch rate (fMb).

Proton range uncertainties

The energy spread of protons, their interactions in non-homogenous tissue, as well as physiological movement of tissues, all contribute to range uncertainties caused by variations in Bragg peak positions. These uncertainties will vary, depending on anatomical site, from millimetres to several centimetres (in the case of physiological movements such as breathing). Patient positioning variations, gain or loss of weight during treatment, with resultant changes in subcutaneous fat depths and tumour shrinkage, can all critically affect particle range, dose and LET distribution, with implications for tumour control and toxicities. Imaging of the Bragg peak position is improving; for example proton nuclear activation of positron emitting isotopes within the body results in PET scanning signal detection to an accuracy of around 2 mm.

A clinical example of some of these considerations, and RBE allocation uncertainties, are shown in figure 2 in Jones (2015) (available on open access). This example illustrates the interdependence of dose, LET and RBE on the potential clinical outcomes. The proton 'dose' shown was calculated using an RBE conversion factor of 1.1 (the equivalent x-ray dose is divided by 1.1). The required 76 Gy dose to the target volume is achieved in both cases, but the proton plan spares the heart, a considerable volume of lung, and breast tissue, so reducing the later risk of breathlessness on exertion, sudden cardiac death and of radiation-induced breast cancer. The small ringed area anterior to the high dose volume shows the position of the spinal cord and the dose received by this structure appears to be same for both treatment plans. Despite these attractive features, what if the RBE of 1.1 used in the dose prescription is higher than that of the tumour and lower than that of the spinal cord? Then, there would be enhanced risks of both tumour recurrence and spinal damage (resulting in paralysis of the lower limbs) when compared with the x-ray based treatment. Also, the effect of increasing particle range, as might happen in a patient with weight loss (so reducing the subcutaneous fat and allowing the beam to travel further), might lead to higher LET and dose in the spinal cord, increasing the risk of paralysis, while also lowering the LET and dose to the tumour, so reducing the probability of tumour control. Weight gain may cause the opposite effects. The simplest allowance for changes in tissue distances would be to allow millimetre exchanges, assuming uniform tissue density. In this way each millimetre lost or gained will shift the high dose boundary in either direction by the same amount. The ballistic advantages, changes in dose distribution and these more uncertain aspects of proton beam therapy (PBT) should be mentioned in informed consent procedures to patients, with explanation of how these risks can be minimised.

To reduce range instabilities, meticulous daily assessment of patient dimensions, accurate tumour localisation, patient immobilisation, etc, including confirmatory measurements made by non-ionising techniques such as ultrasonography, are necessary requirements. Improved dose computing techniques and 4D planning (the three dimensions of space and the fourth of time is included), including adaptive techniques to changes in normal tissues and tumour positions/dimensions with time are even more important in PBT than conventional radiotherapy. Further details of these techniques, including proton radiography, are available elsewhere.

There has always been concern that the highest LET and RBE will occur towards the end of the particle range, even if the Bragg peaks are 'spread out' by the use of range filters or by the spot scanning method. This is because of the greater proportion of high LET radiation in the distal part of the beam; more proximal sections have a lower average LET due to a greater proportion of low LET in the non-plateau section of the SOBPs. The distal end will have the highest LET, which might lead to unintended clinical effects. This concern has led to the intelligent use of 'patch field' techniques, which aligns lateral beam edges (rather than distal edges) against sensitive anatomical structures when using passive beam scattering. In contrast, lateral beam edge beam uncertainties are not so well understood. Beams diverge with depth due to coulomb scattering, and there is geometric divergence with passively scattered beams. Lateral scatter must carry implications for higher LET values, since deviated particles will form Bragg peaks in a more lateral direction. Also, the tracks become more separated and less linear, so a more microvolumetric assessment of energy transfer (MVET) is indicated, as well as the number of particles per unit volume, related to the inter-track distances explained above. There is already evidence that with spot-scanned intensity modulated proton therapy high LET values can be found outside the tumour-bearing target volumes. This is perhaps explained by a greater proportion of partially overlapping beam edges, along with the additional weighting applied to some beamlets, and the smaller inter-track distances.

In the case of carbon ions, the dose profile of each single field (composed of multiple Bragg peaks, as in figure 1.1) was designed to compensate for the inevitable increase in LET along its path (Kanai *et al* 1997): deeper regions, where LET is highest, receive a lower dose; shallower regions are allowed a higher dose because of the lower prevalent LET. Such an elegant arrangement has not been used for proton beams, but should be an urgent consideration, especially when single- or few-field directions are used and if important functional normal tissue is exposed at these distal beam regions. Confirmation of a marked increase in the RBE for lung fibrosis in patients treated by single SOBP proton fields has increased this concern (Underwood 2017). The weighting of proton dose towards the distal end of the beam should be reduced to compensate for such an effect, although the extent will need to be determined by the dose per fraction used, using LET-RBE models, as described in chapters 8 and 9.

1.2 Physics interacting with biology

1.2.1 RBE

The study of the RBE of different radiation qualities is important, since the RBE concept is used in the dose prescription process of hadrontherapy. RBE is defined as:

$$\mathrm{RBE} = \frac{\text{dose of control (low LET) radiation}}{\text{dose of test (higher LET) hadron radiation}}$$

for the same bio-effect, using each class of radiation.

The numerator dose becomes the number by which RBE is scaled from values of one upwards, the denominator being less than the numerator.

The RBE is dependent on the LET of the radiation, with an initial increase in RBE being followed by a reduction at higher LET values due to overkill/or cell killing inefficiency, the turnover point being dependent on the nuclear charge (z) of the ion, but not, apparently, the cell type. The RBE is also inversely related to dose for reasons related to differences in cellular radiosensitivities and the slope of the effectiveness curves between the reference (or control) low LET radiation and the higher LET state. This is illustrated in figure 1.3, based on the linear quadratic model of radiation effect (E) where $E = \alpha d + \beta d^2$, d being the dose, α and β are

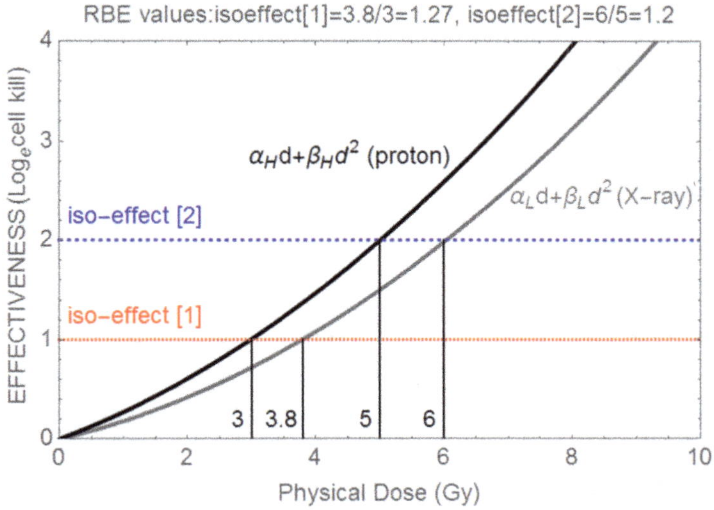

Figure 1.3. Demonstration of the RBE principle for two dose effectiveness curves produced by x-rays (photons) and protons (assuming $\alpha_L = 0.15$ Gy^{-1}, $\alpha_H = 0.24$ Gy^{-1}, $\beta_L = 0.03$ Gy^{-2}, $\beta_H = 0.032$ Gy^{-2}) and where two different iso-effect levels (1 and 2) are considered; their corresponding RBEs are also shown. The numbers 3, 3.8, 5 and 6 refer to the physical doses at which each iso-effect line meets each curve, respectively; d_H followed by d_L for iso-effect 1, d_H and d_L for iso-effect 2. The RBE is given by d_L/d_H in each case, as shown above the figure frame. Reproduced from Jones (2017).

biological coefficients, with subscripts L and H for the low and high LET states; further details regarding the radiobiology are given in following chapters. The RBE falls from 1.27 to 1.2 due to the geometry of the curves and such a change is an inherent feature of the linear-quadratic model providing that the LET-induced increment in α exceeds that in β (further discussed in chapters 8 and 9).

The RBE concept was included in the dose prescription system for fast neutron therapy and continues to be important in proton and ion beam therapy; it is also potentially relevant in nuclear medicine and some forms of brachytherapy, but has not yet been used in their clinical application.

In experiments to determine the RBE, the choice of the control (or reference) radiation has varied considerably; examples include low-voltage x-rays (100–200 keV peak) and orthovoltage (200–350 keV peak), along with variable degrees of filtration to remove the lowermost keV x-rays within the x-ray spectrum. Other examples include 137-caesium or 60-Co gamma rays, megavoltage electrons and, more rarely, megavoltage x-rays in the clinical range 4–10 MeV. The lower the energy of any beam, the greater the LET; this applies to x- and γ-rays, protons, and all ions and electrons, because lower energies result in more localised energy release and absorption, and a higher LET, and also results in more clustered DNA damage. This will increase the RBE up to a saturation level beyond which the RBE falls at even higher LET values. Conversely, at higher energies of a particle or photon beam, then local LET and RBE are reduced, because the ionisations are not so spatially clustered.

Since the reference radiation (sometimes called the control radiation) supplies the numerator dose for the RBE estimation, this must exceed the hadron radiation dose for the same bio-effect in order for the RBE to be greater than one. Should the control RBE irradiator have an LET that is greater than the test hadron LET then the RBE will be less than one. Choosing a control irradiator that has a higher LET than that of the photons/x-rays used in hospital clinics, will give a lower estimate of RBE than would be the case in the hospital setting for the same cellular system.

1.2.2 Choice of the control (or reference) radiation source

It follows that the choice of the reference radiation will have an important effect on the RBE ratio found. For example Hall and Gaccia (2011) quote typical LET values (keV μm^{-1}) of 2 for 250 keV (but the filtration is unspecified), 0.2 for 60-Co, 4.7 for 10 MeV protons and 0.5 for 150 MeV protons, while megavoltage photons are not mentioned. More comprehensive, but still incomplete, data can be found in ICRU report 16 (1970). These suggest, by extrapolation, that megavoltage photons in the clinical range of 4–8 MeV would have a LET value of around 0.2 keV μm^{-1}. A useful collation of data in Andrews (1978) shows a mean LET of 0.5 keV μm^{-1} for 3 MeV photons and 60-cobalt irradiation, but stopping powers (or LET_∞) of 0.18 for 2 MeV electrons as well as being the theoretical minimum for any charged particle, and hence an important limit or baseline for relative biological effect studies. The average Compton electrons produced from 60-cobalt γ-rays have an LET_∞ of 0.26 keV μm^{-1}, so a slightly lower value around 0.2 keV μm^{-1} would be expected from clinical range megavoltage photons.

Further information is contained in some publications, such as Spadinger and Palcic (1992), where an LET range of 5.5–6 keV μm^{-1} is quoted for 25 keV x-rays and 1.7 keV μm^{-1} for 122 KeV x-rays (the average value obtained with 250 KeV x-rays with 0.35 mm Cu and 0.4 mm Sn filtration), with a value of 0.2 keV μm^{-1} for 11 MeV electrons, which are likely to closely resemble megavoltage photons in the clinical range.

From the above data, it seems reasonable to use a control radiation of either electrons of photons, according to the experiment being done, with an LET value of around 0.2 keV μm^{-1}, as would be achieved with a clinical photon or electron beam. The added value of having a megavoltage photon beam would be for comparative combined ballistic and radiobiological studies in humanoid phantoms between charged particles and clinically relevant photon energies.

Some important examples of the unsuitability of orthovoltage and lower voltage x-rays beams as the control/reference radiation are as follows.

1. For the collated proton RBE data of Paganetti *et al* (2002), orthovoltage radiations, 14 low voltage or orthovoltage x-ray irradiations had an RBE of less than one and only one such experiment showed an RBE of just over one. The probability of this being due to chance or experimental error is low; use of the sign test provides the probability of there being no difference as $p = 0.00061$, a highly significant result.

2. The literature contains a report by Amols *et al* (1986) who measured cell survival data in DLD-1 human tumour cells. Their results clearly demonstrate a statistically significant RBE difference between orthovoltage and megavoltage radiation ($p = 0.001$). A small difference was also measured in the RBE between megavoltage photons and megavoltage electrons, but the difference was not statistically significant ($p = 0.25$). All biological, dosimetric, and microdosimetric data were obtained under nearly identical geometric conditions, but using only one cell line.

The past literature contains many contributions that concentrate on the influence of lower photon energies in the low kilovoltage range on the RBE, but these are mainly in the context of radiation protection, because this class of x-rays are used so much in routine diagnostic medicine. Such work has confirmed that the RBE varies inversely with photon and particle energy (see, for example, the review by Hill 2004).

The RBE should be taken into account in clinical radiotherapy using positively charged particle beams, where RBEs can be significant even in the case of protons. Papers by Belli *et al* (2000), Britten *et al* (2013) and Marshall *et al* (2016), are examples of such work, containing well-documented experimental detail. There is presently a potential shift of opinion to include more flexible (or variable) RBE values than the constant values previously assumed in proton therapy, in order not to: (a) under-dose certain radiosensitive tumour classes and (b) over-dose late-reacting normal tissues of clinical importance. This is discussed in later chapters.

The urgent need to clarify RBEs in critical experimental normal tissues exposed to proton and other ionic beams will require careful choice of control radiation if the purpose of such studies is to provide information that will ultimately be useful for clinical purposes. Some potential forms of control radiations can easily be eliminated, as is the case for orthovoltage x-rays, despite their low expense and convenience, should be avoided, since they will lead to lower estimates of the RBE. 60-cobalt units also have some significant disadvantages. These include lower dose rates with time, since it is advisable to use the same exposure times (within differences of only 1–2 min at the most) for the control and test irradiations, due to ongoing repair of sub-lethal radiation damage. Longer treatments will have slightly lower bio-effectiveness and this will influence the estimated RBE ratio. Also, the need to compensate for radioactive decay on a weekly or monthly basis may sometimes be forgotten or wrongly applied; the LET can be lower than for megavoltage photons and there is drift of dose rate with time, and modification of source treatment distances to compensate for radiation decay, with accompanying changes in depth dose distributions. There are also problems associated with radiation protection of staff, shielding requirements, and source replacements required with time, are expensive, and have their own difficulties.

Some laboratories do have a 137-caesium irradiator, which is more compact, and delivers monoenergetic photons close to 0.5 MeV, which should have an LET close to that of cobalt beams. Again radioactive decay has to be compensated for and dose rate effects may accrue with time, although this can be overcome by using shorter source to target distances. Such a unit can be fitted into the size occupied by a large double refrigerator without additional shielding.

Megavoltage electrons are a reasonable choice for the control radiation in RBE studies. However, electron dosimetry, although much improved in recent years, has limitations due to its high scatter, especially for small field sizes, as well as an energy-dependent relationship of electron build-up with depth, which must be taken into account in any laboratory experiment. Electron fields are essentially SOBPs, often with lower RBEs than fields in megavoltage electron plateau regions, but there remains uncertainty as to their higher RBE values with fall off of dose and energy towards their end of range. It must not be forgotten that Auger electrons (from low energy radioactive emissions) can have very high RBEs. However a broad electron beam operating at, say, 8–10 MeV would give a satisfactory uniform dose for radiobiological experiments, but could not be used for beam ballistics combined with radiobiology comparisons.

There is interest in some particle therapy centres across Europe, including international laboratories like CERN, in addressing the clinically important RBE issue. For comparable results it is very important that the control radiation be standardised. The best standard to choose, despite the additional expense, would undoubtedly be megavoltage x-rays from a linear accelerator operating at typical energies used in the clinic. It is already the case that national radiation standards laboratories such NPL (Teddington, UK) contain a standard hospital linear accelerator for dosimetry comparisons and maintenance of standards. Likewise, laboratories that aim to contribute useful RBE data should use a control megavoltage linear accelerator with optional use of 4–10 MeV photons or electrons for their studies. A single 137-caesium source could also be useful.

Interestingly, since hospital linear accelerators are normally replaced every 10 years, it is relatively easy to obtain one that can be refurbished for experimental use. These would have a gantry for variable beam angles and collimation down to small field sizes.

1.2.3 Does the RBE reduce with depth in scattered beams?

In general, the RBE increases with depth due to the progressive increase in LET, especially within the Bragg peaks. However, there is evidence that if the same LET and dose is given at the same SOBP positions situated at markedly different depths, achieved by using a different incident energy, then the RBE is reduced in the deepest position. These intriguing experiments, led by V Megnin-Chanet (Institute Curie), were designed to study proton RBEs at various depths using the Orsay proton beams (Calugaru *et al* 2011), with two different incident energies of 76 and 201 MeV, with mid SOBPs at 20.5 and 160 mm, respectively, and also near the distal end of SOBPs at 26.9 and 185 mm, respectively, with identical LET characteristics towards their distal ends. The main physical differences appeared to be the beam widths: 3 and 12 cm diameters, respectively, indicating greater lateral dispersion at the increased depth after passive beam scattering. They observed a reduction in the RBE in two human cell lines from around the highest RBEs of 1.3–1.33 (at 3 cm depth) down to

1.0 (at 12 cm depth). Although the dose rates did differ with increasing depth, when DNA repair kinetics are considered the changes in minutes required to deliver the dose ranges used cannot account for the change in the RBE (the dose rate effect is discussed further in chapter 2).

In contrast, Britten *et al* (2013) found no such effect in two cell lines with incident energies of 87 and 200 MeV at Bloomington, where scanned beams were used. Scanned beams, without metallic filter-related scatter, essentially consist of parallel tracks and so deliver the shorter inter-track separation distance (s) over most of their range. They exhibit little geometric dispersion, so s is relatively well preserved with depth, only being affected by mild coulombic scattering, which manifests itself mostly near to the extreme range when energy and velocity are reduced. Alternative explanations for the findings of the Paris and Bloomington experiments might include chance, or a cellular batch effect, or the cell types used. These latter hypotheses are unlikely, as each Institute had used two different cell lines with different radiobiological properties.

The most obvious physical explanation of this phenomenon, although speculative, is the increasing track separation caused by divergence in passively scattered beams. There can also be an associated loss of RBE with depth owing to increasing γ-ray contributions, but these are small and would be at least partly offset by neutron production in matter, and this should be much the same in both centres. The geometric scattering from a metallic scattering foil will result in ISL effects, emerging from an apparent point source, so that radiation fluence will fall sharply with increasing distance or depth.

If it is assumed that the RBE is proportional to inter-track distance, which decreases with depth (x) in passively scattered beams, all other factors being equal (dose, energy, LET), and if the rate of change of this effect depends on the operative RBE, then

$$d(\Delta\text{RBE})/dx = -K\Delta\text{RBE},$$

where ΔRBE represents the RBE value in excess of 1.

Integration then provides, for RBE values >1, $1 + \text{RBE}_x = 1 + \text{RBE}_0\, e^{-cx}$ where RBE_x is the RBE value in excess of 1 at a specified depth of x cm, RBE_0 is the RBE value in excess of 1 at the smallest depth, and c is a constant assumed to be controlled by track separation.

For the Megnin-Chanet experiments, and using their pooled data for two different human cell lines, the above function provides a value of $c = 0.009$ and 0.013, respectively, for the mid SOBP and distal SOBP positions used for the comparisons, and so it is possible to plot the reduction in the RBE with increasing depth, as shown in figure 1.4. Further experiments are required at multiple depths to confirm this hypothesis and achieve a more robust value for parameter c.

The other approach might be to use the inter-track distance (s). Using similar triangles, the radius of a divergent overall beam-field is proportional to the distance (x) that it traverses. For two different depth positions 1 and 2, indicated by the respective subscripts, the overall beam radii r_1 and r_2 are proportional to their respective depths (x_1 and x_2) due to the similar triangle rule; the areas A_1 and A_2 are

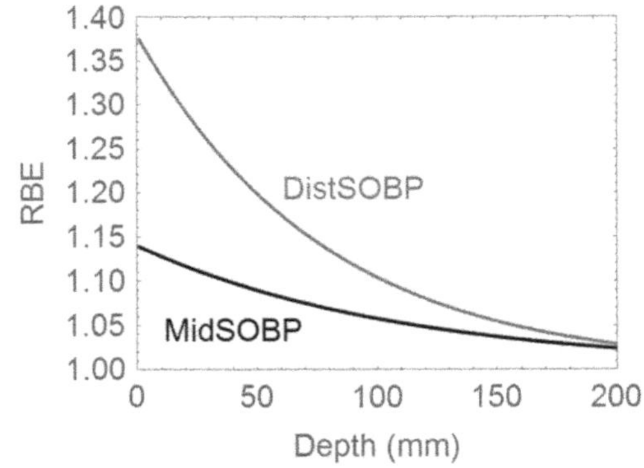

Figure 1.4. Putative relationship between the RBE and depth at the mid SOBP and distal SOBP obtained from data averaged for two human cell lines (HeLa and a head and neck squamous cell cancer line) for caesium and cobalt referenced RBE values obtained using the above equations.

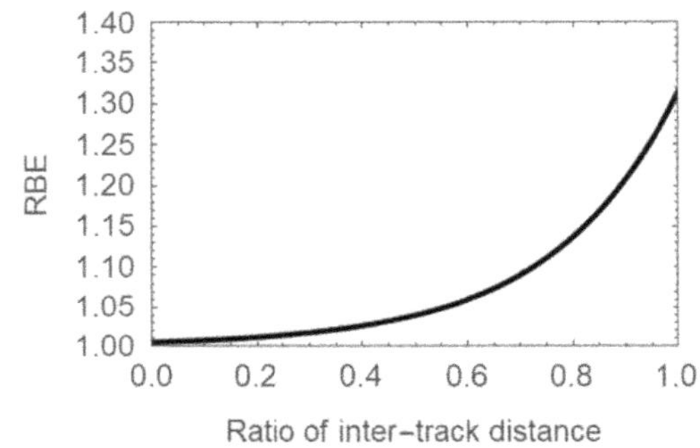

Figure 1.5. Plot of the RBE against the inter-track distance ratio (SR), normalised to $s = 1$ where the RBE is 1.33, but with an RBE close to 1 when the SR is around 0.1 for scattered proton beams, as in Calugaru *et al* (2011).

given by πr_1^2 and πr_2^2. The inter-track distance, when squared, is $s^2 = A/N$, and since N remains constant, then the ratio of inter-track distances (s_1/s_2) at two different depths will be the same as x_1/x_2. Using one of the data examples in Calugaru *et al* 2011, from $x = 2.69$ to 18.5 cm, the inter-track separation ratio (SR) will change from the shallowest depth, where s_1 is normalised to 1, down to $2.69/18.5 = 0.145$ (in the absence of precise knowledge of N or F). When this process is applied to the data, and by using a similar exponential function as in the above example, then the relationship between the RBE and the inter-track separation ratio (SR) can be expressed as $1 + 0.0055\ e^{(4.14\text{SR})}$, and this function is plotted in figure 1.5.

Such exploratory data needs to be extended, with RBE estimations at many depths in a panel of cell lines in order to fully verify and extend this hypothesis, using further microdosimetry considerations.

Although speculative, if such a reduction in RBE with depth can occur in some cell lines with divergent beams, it may be possible to investigate reducing potential RBE problems in normal tissue adjacent to a tumour by electively increasing the path length (and energy) of the radiation and adding tissue equivalent material

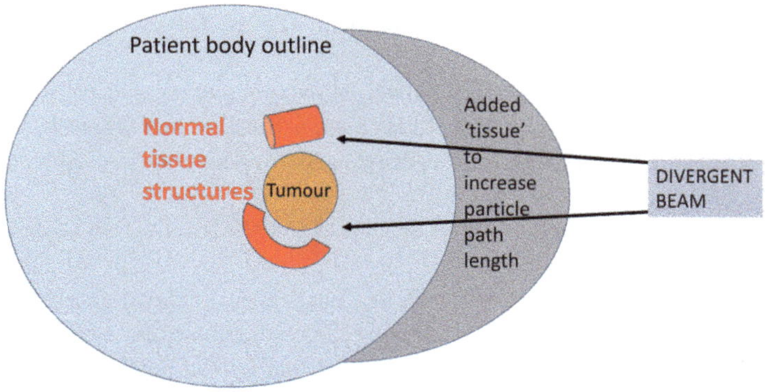

Figure 1.6. Potential treatment set up for deliberate reduction of RBE with depth using a longer treatment track distance by adding artificial 'tissue'.

outside the skin (as shown in figure 1.6), aiming for the RBE to be close to one at the critical distance. The length of additional material would require careful calculation to produce the SOBP at the required tissue depth, but with loss of the RBE this could be advantageous where the local biological characteristics confer a high normal tissue RBE in a critical structure and which exceeds any critical normal tissue dose sparing advantage. The alternative way of reducing the normal tissue RBE is to increase the dose per fraction of the particle therapy (these aspects are discussed in later chapters).

The possible detriment would be loss of electronic equilibrium build-up in skin and subcutaneous tissue due to the introduction of the artificial tissue, which allows the beam to diverge to the required extent to reduce the RBE.

References

Andrews J R 1978 *The Radiobiology of Human Cancer Radiotherapy* (Baltimore, MA: University Park) pp 241–4

Amols H I, Lagueux B and Cagna D 1986 Radiobiological effectiveness (RBE) of megavoltage x-ray and electron beams in radiotherapy *Radiat. Res.* **105** 58–67

Belli M *et al* 2000 Inactivation of human normal and tumour cells irradiated with low energy protons *Int. J. Radiat. Biol.* **76** 831–9

Britten R A *et al* 2013 Variations in the RBE for cell killing along the depth-dose profile of a modulated proton therapy beam *Radiat. Res.* **179** 21–8

Calugaru V, Nauraye C, Noël G, Giocanti N, Favaudon V and Mégnin-Chanet F 2011 Radiobiological characterization of two therapeutic proton beams with different initial energy spectra used at the Institut Curie Proton Therapy Center in Orsay. *Int. J. Radiat. Oncol. Biol. Phys.* **81** 1136–43

Hall E J and Giaccia A J 2012 *Radiobiology for the Radiologist* 7th edn (Philadelphia, PA: Kluwer/Lippincott Williams) p 106

Hill M A 2004 The variation in biological effectiveness of x-rays and gamma rays with energy *Radiat. Prot. Dosimetry.* **112** 471–81

ICRU Report 16 1970 *Linear Energy Transfer* (Bethesda, MD: ICRU)

Jones B 2015 Towards achieving the full clinical potential of proton therapy by inclusion of LET and RBE models *Cancers* **7** 460–80

Jones B 2017 Proton radiobiology and its clinical implications ecancer *ecancer* **11** 777

Kanai T, Furusawa Y, Fukutsu K, Itsukaichi H, Eguchi-Kasai K and Ohara H 1997 Irradiation of mixed beam and design of spread-out Bragg peak for heavy-ion radiotherapy *Radiat. Res.* **147** 78–85

Katz R, Ackerson B, Homoyooufar M and Sharma S C 1971 Inactivation of cells by heavy ion bombardment *Radiat. Res.* **47** 402–25

Marshall T I, Chaudhary P, Michaelidesová A, Vachelová J, Davídková M, Vondráček V, Schettino G and Prise K M 2016 Investigating the implications of a variable RBE on proton dose fractionation across a clinical pencil beam scanned spread-out Bragg peak *Int. J. Radiat. Oncol. Biol. Phys.* **95** 70–7

Newhauser W D and Zhang R 2015 The physics of proton therapy *Phys. Med. Biol.* **21** 155–209

Paganetti H, Niemierko A, Ancukiewicz M, Gerweck Le, Goitein M, Loeffler J S and Suit H D 2002 Relative biological effectiveness (RBE) values for proton beam therapy *Int. J. Radiat. Oncol. Biol. Phys.* **53** 407–21

Peach K, Wilson P and Jones B 2011 Accelerator science in medical physics *Brit. J. Radiol.* **84** S1–S10

Spadinger I and Palcic B 1992 The relative biological effectiveness of 60Co gamma-rays, 55 kVp x-rays, 250 kVp x-rays, and 11 MeV electrons at low doses *Int. J. Radiat. Biol.* **61** 345–53

Underwood T 2017 Oral presentation *Scott-Fowler Symposium, Oxford*, September 2017

Wilson E J N 2001 *An Introduction to Particle Accelerators* (Oxford: Clarendon)

Chapter 2

The essential radiobiology background

2.1 Introduction

The information given is intended as a summary for those with existing basic knowledge. It is divided into two sections: the first describes the models and their background: the second considers the choice of some key parameters.

2.2 Background and models

2.2.1 The linear-quadratic (LQ) model

Douglas Lea, a pupil of Rutherford, fitted the average yield of severe (or lethal) chromosomal aberrations per cell (E) in the form of $E = \alpha d + \beta d^2$ (Lea 1962). For many years, this finding was largely ignored because of the increasing use of target theory models that had little biological basis. The LQ model later became important for two reasons.

Firstly, several mathematically complex theoretical approaches, separately based on microdosimetry, molecular damage (and its repairability/non-repairability) and chromosomal aberration yields, all approximate to the LQ formulation, as in Rossi and Zaider (1992), Curtis (1986), Chadwick and Leenhouts (1973), and Campbell and Warenius (1989).

Secondly, animal and human fractionation data were fitted by transformations of the LQ model. Douglas and Fowler (1976) used fractionation effect (or Fe) plots. These used a simple algebraic transformation of the LQ as $1/D = \alpha/E + (\beta/E)d$, where D is the product of each 'fractional' dose d multiplied by the number of fractions n. The reciprocal of total dose when plotted against dose per fraction produced reasonably good linear fits of iso-effective fractionation data, where the same biological endpoint was obtained using different dose fractionation schedules. These plots were used to obtain the α/β ratio by taking the intercept divided by the slope. Thames *et al* (1982), using multiple iso-effective data sets, later showed the very important and clear dissociation between classes of tissues with different α/β

doi:10.1088/978-0-7503-1338-4ch2

ratios by plotting dose per fraction against total dose: acute reacting tissues (rapidly growing biosystems, such as rapidly growing tissues and normal epithelial tissues) had high α/β ratios (7–15 Gy), but late reacting tissues (slowly growing tumours or low turnover systems including all stable normal tissues) had low α/β ratios (2–3 Gy). In comparison to the non-biologically based power law models, which failed at high and low dose limits, and were initially judged to apply to all tissues and tumours, the LQ approach was more flexible and produced more realistic predictions that could be applied over a wide range of doses.

Critics of the LQ model appear to be fixed on its simplicity and assume that it relies on the simplistic textbook diagram Keller and Rossi approach, where the DNA double helix is used to demonstrate single hit and double hit lesions; such a concept is naïve, in that the LQ endpoint must be the yield (or probability) of lethal events per cell or even across a whole tissue, so that α and β must represent the ultimate coefficients of lethality or clinical effects, rather than some of their initial DNA damage precursors. Criticism has rightly been raised as to how such a simple formulation can represent the entire pathway of injury from a cellular to the tissue level, where many biological processes (inflammatory mediators, cytokines and cellular sub-compartmental kinetics, fibrosis, vascular damage, and tissue hypoxia and nutrient depletion) may all influence the end result. However, even complex industrial processes, as well as cellular processes, are governed by key rate-limiting steps, which can simplify any analysis of outcomes.

The present book considers high linear energy transfer (LET) effects from protons (and other ions). It is useful to consider the dimensions of LET (energy released per µm of track). Micrometres are a more appropriate measurement for chromosomal damage (resulting from multiple smaller DNA changes, but only where these accumulate in a critical 3D space to produce lethal chromosomal breaks. The chromosomal diameter is around 700 nm, and each non-replicating chromosome is covered by heterochromatin protein, which can also be chemically disrupted by the clustered radiation in high LET beams, releasing native damaged DNA to the oxidative micro-environment of the cell, and disrupting any on-going background repair mechanisms. Consequently, as LET increases, the complexity of local chemical disruption to DNA becomes beyond the ability of the cell to repair the cumulative local damage—resulting in a lethal chromosomal break.

A good example of LQ model clinical utility is given in figure 2.1, where the LQ model has been used to delineate an iso-effective dose curve for spinal cord myelitis (which can cause paralysis), based on the low risk associated with a dose of 50 Gy in 25 fractions, with data points given by many empirically derived clinical fractionation schedules used in the clinic by the present author, and considered to be very safe (grey points), whereas the doses used to treat common cancers are indicated by the red points. The curve provides an effective upper limit, which should not be exceeded in the spinal cord tissue if uniformly irradiated in cross section. Details of how to construct such a curve, using the biological effective dose concept, are given in later sections.

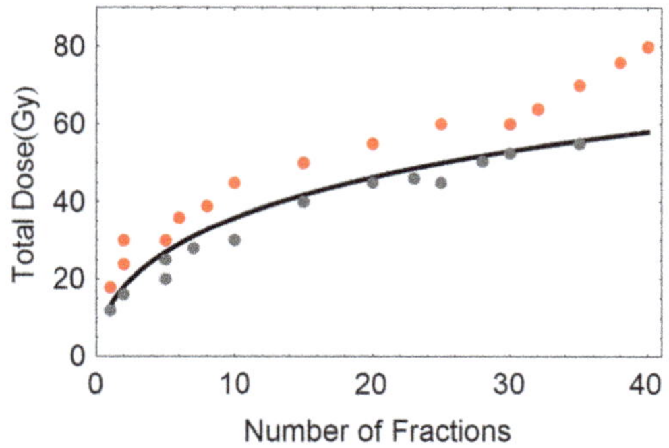

Figure 2.1. LQ model-based plot of the relationship between total dose and number of fractions, which results in the same low risk (iso-effect) of spinal cord paralysis, obtained using an α/β ratio of 2 Gy. The grey points are considered to be safe. The red points are the doses often used to treat squamous cell cancers and which exceed the iso-effect curve. Clinical radiotherapy may utilise the grey dose-fractionation points to palliate cancer close to the spinal cord, but when the red point curative doses are given, great care is taken to ensure that the spinal cord doses are either on or below the curve.

A more detailed summary of the LQ model is given in box 2.1.

Box 2.1: Summary of the LQ model

Contrary to many statements in textbooks, and common misconceptions, α and β are *not* specific coefficients for single strand (SSBs) and double strand (DSBs) DNA breaks, the yields of which are dose dependent and they are in most instances efficiently repaired by cellular enzymatic mechanisms. A large number of SSBs and DSBs are required to exist in reasonably close proximity, and avoid repair in order to produce a lethal event, normally accompanied by the appearance of a lethal chromosomal aberration, which will cause cell death due to asymmetric segregation of genetic information, usually at the next mitosis. Consequently SSBs and DSBs are necessary precursors, but not sufficient requirements, for lethality; additional spatial and temporal considerations may or may not lead to lethality. Consequently, in cell survival experiments, α and β must reflect the probability of lethality with increasing dose and are best defined as:

α is the coefficient of lethal events per cell per unit dose (units Gy^{-1}),

β is the coefficient of lethal events per cell per unit dose squared (units Gy^{-2}).

In this way, α and β probably reflect the number of lethal chromosomal events with dose and dose squared, respectively.

It follows that the probability of cell survival (where there are no lethal events), referred to as the survival fraction (SF), is obtained using Poisson statistics:

$$\mathrm{SF} = \mathrm{e}^{-(\text{expected number of lethal events})} = \mathrm{e}^{-ad-bd^2}.$$

Summary of events

DNA level events: base changes, SSBs, DSBs, short patch, long patch. Cellular response systems: repair by multiple mechanisms, but misrepair or inability to completely repair results in lethality according to local complexity. These, if sufficiently concentrated in a particular location on a chromosome or chromatid will lead to:

chromosomal and chromatid level events caused by asymmetrical breakage at stressed sections (thus clustered radiation damage will cause more mechanical stress). Cellular responses: time dependent repair of DNA damage, and attempted re-joining resulting, minimally, in some loss of heterozygosity; more severe and multiple breaks result in lethality. The numbers of lethal events per cell are expressed as α and β per unit dose and dose squared, respectively.

The LQ can alternatively be expressed as:

$\alpha = \Sigma$ (localised unrepaired DNA damage $\Rightarrow$ lethal chromosomal injury events per cell)/dose.

$\beta = \Sigma$ (localised unrepaired DNA damage $\Rightarrow$ lethal chromosomal injury events per cell)/dose^2.

With a greater the degree of local clustering of ionisation (as with high LET) then α and β are expected to increase.

The predominant mode of cell killing at low dose is via α-mediated damage, especially direct damage (less sensitive to oxygenation status), invoking a repair mechanism called non-homologous end-joining (NHEJ) operating on strand breaks throughout the cell cycle, and is highly susceptible to increased LET (Takahashi *et al* 2014); the damage is cumulative with dose and not dependent on further dose commitments, so the damage caused will not depend on the dose rate.

The other mode of cell killing, which dominates at high dose, is via β-mediated damage, especially indirect damage (more sensitive to oxygenation status or sensitising drugs), inter-track cooperation may occur, and a superior repair mechanism—recombination repair (RR) operates by recombination and chromatid exchanges from non-damaged DNA regions. This form of cell killing is less susceptible to increased LET, and exhibits a pronounced dose rate effect because of the requirement for two separate dose related events, which lead to the squared term.

Note that experimental lethality coefficient α and β values, as determined in cell survival experiments or from *in vivo* tissue work, will also include other cell killing processes such as:

- apoptotic cell death (saturates at low dose), so is a fixed proportion at clinically used doses above 1.5 Gy,
- autophagy (dose),
- bystander effects (again occurs mainly at low dose).

These three mechanisms are explained further in radiobiology textbooks.

2.2.1.1 Model variants

Clinicians especially should be aware of two classes of models that can assist with the relative biological effect (RBE) determination, and therefore fractionation issues.

Bottom-up models. Examples include the developing GEANT4-DNA project, and detailed micro-dosimetry studies, such as the local effect model (Kase *et al* 2008) based on LET values, together with assumed or measured (the so-called low-level) sub-cellular and cellular parameters, which allow the RBE to be predicted, although the model remains to be perfected. This and other variants can be used to guide the choice of dose and so aid treatment planning by application on a voxel to voxel basis, etc. These are discussed further in chapters 8 and 9.

Top-down models. These use high level parameters (at the clinical level) such as the biological effective dose (BED), tissue or tumour α/β ratios, or the α and β parameters themselves, and will be discussed below and in later chapters. They link our present knowledge of clinical radiobiology with fractionation phenomena and can usually be applied with ease using relatively simple mathematics for low and high LET radiations. They are essentially phenomenological, but with a real biological basis. They are an effective check for the clinician, who should ideally be able to perform such calculations, and understand their implications and limitations.

Obtaining a model that contains the multi-scaled nature of the events must remain a future goal, although it may not be fully obtainable because it is not possible to define the event probabilities down to individual cells or indeed know all the contributing parameters, the heterogeneity of cell sensitivities, which vary with position in the cell reproductive cycle, the biochemical red-ox state, the nuclear and cell volumes, repair proficiency, etc. This type of challenge exists in many parts of science, including physics where the separate quantum and gravitational models have yet to be unified in a single framework, despite each model being highly accurate in their own domains. However, there are some realistic assessments of the scaling of the RBE from the low LET α/β ratio in some micro-dosimetry related models as in Japan and Germany (Elsässer and Scholz 2007, Kase *et al* 2008), for example to predict brain RBE using the preferred α/β of 2 Gy, and further simpler options are given in chapters 8 and 9.

2.2.1.2 BED

Barendsen (1982), Dale (1985, 1989) and Fowler (1989), all contributed to the application of the BED concept to clinical and experimental data sets. In earlier publications the term extrapolated response dose was used, but this is synonymous with the BED.

For standard megavoltage low LET radiation the following approach is taken. From the surviving fraction (SF) of cells after radiation exposure, included in a Poissonian probability function for the probability of no lethal damage (i.e. survival) when the expected number per cell is E, consequently

$$\mathrm{SF} = \mathrm{e}^{-(E)} = \mathrm{e}^{-\alpha d - bd^2}, \tag{2.1}$$

and for n fractions it follows that

$$\mathrm{SF}^n = \mathrm{e}^{-n(\alpha d + bd^2)}. \tag{2.2}$$

By taking the natural logarithm and multiplying by −1, and replacing $-\ln(\mathrm{SF}^n)$ by E (sometimes called the 'log cell kill') a much simpler expression is obtained:

$$E = n(\alpha d + \beta d^2). \tag{2.3}$$

Next consider the limit of low dose, where the βd^2 term can be ignored. Then, E approaches $n\alpha d$ as $d \to 0$, and so $E/\alpha = nd$ at near zero dose. The nd product is the total dose (D) required for a given effect using a large number of very small dose fractions, and represents a ceiling of dose. At very low dose, this is the BED, expressed in units of Gy. In this way the BED can be defined as the total dose required for a specific bio-effect, if the dose is delivered in a very large number of fractions of low dose.

Returning to the standard LQ formula (equation (2.1)), then division throughout by α leads to the important expression for BED as:

$$\mathrm{BED} = nd\left(1 + \frac{d}{\frac{\alpha}{\beta}}\right). \tag{2.4}$$

This applies to low LET radiation, noting that the first term in the parentheses is one.

For any specified BED, the equivalent dose in 2 Gy fractions (EQD_2) is simply:

$$\mathrm{EQD}_2 = \mathrm{BED}/(1 + 2/(\alpha/\beta)). \tag{2.5}$$

It must be noted that the BED, expressed as $\mathrm{Gy}_{[x]}$, should be further qualified by x, which is the specific α/β ratio used.

2.2.1.3 Repopulation allowances

BED modifications are required to account for ongoing tumour cell repopulation during treatment. Since tumour cells grow exponentially in good growth conditions, it is now standard to express this process as a reduction of the BED by using a BED-dose equivalent of repopulation, obtained in the following way.

Repopulation effects may be incorporated into the LQ model. The SF following radiotherapy is given by equation (2.1) above, but the surviving clonogenic cells will repopulate independently of the radiotherapy, so that SF must be increased by a factor that accounts for the exponential cellular repopulation during treatment given over a time t, so the equation becomes:

$$S = \mathrm{e}^{-n(\alpha d+\beta d^2)} \cdot \mathrm{e}^{\frac{\ln 2.t}{T_{\mathrm{eff}}}} = \mathrm{e}^{-n(\alpha d+\beta d^2)+\frac{\ln 2.t}{T_{\mathrm{eff}}}} \tag{2.6}$$

where T_{eff} expresses the effective doubling time of the repopulating cells, which may vary during treatment.

By taking natural logs of each side of the equation, the logarithmic surviving fraction expressed as the negative natural logarithm (or radiation log cell kill, E_t) at time t is given by:

$$-\ln S = E_t = n(\alpha d + \beta d) - 0.693\frac{t}{T_{\mathrm{eff}}}. \tag{2.7}$$

The sign of the repopulation factor is now negative, and the cell kill part of the equation is positive for consistency.

There has been much debate regarding the form of the repopulation factor and the nomenclature of the doubling time. It is reasonable at this stage to distinguish between the actual clonogen cellular doubling time, T_{C}, the potential doubling time, T_{pot}, and the effective doubling time denoted by T_{eff}. T_{eff} is a rather flexible parameter in that it may initially correspond to the pre-treatment tumour volume doubling time, T_D, but in some tumours may shorten during treatment and approach the value of the pretreatment T_{pot}.

For normal tissues, no allowance for repopulation should normally be necessary for repopulation during treatment in the case of late responding tissues. Inclusion of normal tissue repopulation, or 'recovery', following treatment may be required when considering re-treatment after an interval of many months or years. This is discussed further in chapter 9. There is currently no general agreement on how to include a variable repopulation factor for acute tissue effects, where α/β ratios will be larger and appear to increase during the treatment from average values of 4, 12 and 35 Gy in the first 2, 2–4 and 4–6 weeks, respectively, of treatment as far as the skin is concerned (Hopewell *et al* 2003).

2.2.1.4 BED and repopulation

To allow for the effect of repopulation during treatment the standard BED equation requires a negative repopulation factor that here assumes a constant rate of BED compensation during treatment:

$$\mathrm{BED} = D\left[1 + \frac{d}{\alpha/\beta}\right] - KT \tag{2.8}$$

where K (units of Gy per day) is the daily BED equivalent of repopulation and T is the overall treatment duration. Typical values of K used for squamous cell cancers are 0.5–1.0 $\mathrm{Gy_{10}}$ $\mathrm{d^{-1}}$. These are considered to be average values throughout treatment. It is emphasised that K values are in units of physical dose, and do not normally represent the physical dose per day required to offset repopulation. This important difference can lead to errors in calculating compensatory schedules for unintentional treatment interruptions, as shown by Dale *et al* (2002), and the situation is more complex for high LET treatments (see chapter 9). For slower growing tumours, e.g. breast and prostate, there is little evidence for repopulation until durations of seven weeks or more have elapsed when the K value may be close to 0.2–0.3 Gy $\mathrm{d^{-1}}$ (Thames *et al* 2010). In the case of very anaplastic or poorly differentiated tumours with total lack of growth control mechanisms it is possible that repopulation will be continuous without a lag-time. The term K includes several parameters; it is obtained by dividing the conventional repopulation correction factor by the α parameter (just as BED is obtained by dividing the log cell kill (E) by α, so that

$$K = \frac{0.693 \cdot t}{\alpha \cdot T_{\mathrm{eff}}}. \tag{2.9}$$

Since α increases with increasing LET, so the K value reduces and the dose to compensate for unintended treatment interruptions will change compared to the dose required for megavoltage photons. These subtleties are further discussed in chapter 10.

Alternative patterns of repopulation

The above formulation assumes *continuous* repopulation (i.e. the repopulation rate is continuous and at the same rate during the entire treatment time). For tumours where steady state repopulation begins after an apparent time lag of T_K days, the repopulation term is modified to be $K(T - T_K)$. The BED is then:

$$\text{BED} = nd(1 + d/(\alpha/\beta)) - K(T - T_K). \tag{2.10}$$

This pattern of repopulation can be described as being *discontinuous*. For head and neck cancers, T_K is between 21 and 28 days. This approach is only valid if $T > T_K$. It would be incorrect to assume that all tumours behave like most differentiated squamous cell cancers. For ultra-short fractionation schedules with overall times shorter than 21 days the repopulation correction factor can either be omitted, as there is relatively little repopulation, or it is possible to assume that a longer T_{eff} is operational, for example assume that T_{eff} is 5–10 times larger, especially if the tumour is poorly differentiated. An alternative formulation, based on a *progressive* or *delayed* reduction in the cell loss factor during treatment can also be used. This model is based on a continuously changing repopulation rate, governed initially by the volume doubling time. Later during treatment, when ϕ values are small, T_{eff} approximates to the pre-treatment T_{pot} value. This model predicts a slowly increasing extra dose required to counteract repopulation as treatment time is increased, which simulates accelerated repopulation, and there is the attractive feature of no sudden discontinuities. A wider discussion and alternative approaches are given by Jones and Dale (2007).

2.2.1.5 BED expression of high LET radiation

For higher LET radiations it is necessary to include the separate increments in α and β due to increasing LET, noting that the former normally exceeds the latter, as in Jones *et al* (2006b), Carabe-Fernandez *et al* (2007). Some authors ignore the smaller β increments for low dose treatments (see chapters 7 and 8), but the β contribution to effectiveness becomes progressively more important with higher dose per fraction (often referred to as hypofractioned radiotherapy).

First, it is necessary to apply basic limit theory to the basic LQ model in the form of the following iso-effective equation (where the same bio-effect is produced by two forms of radiation in each case designated by subscripts L and H for low LET and high LET, respectively, as:

$$n_{\text{L}}\left(\alpha_{\text{L}} d_{\text{L}} + \beta_{\text{L}} d_{\text{L}}^2\right) = n_{\text{H}}\left(\alpha_{\text{H}} d + \beta_{\text{H}} d_{\text{H}}^2\right). \tag{2.11}$$

To simplify further, assume that $n_{\text{L}} = n_{\text{H}}$, so that

$$\alpha_{\text{L}} d_{\text{L}} + \beta_{\text{L}} d_{\text{L}}^2 = \alpha_{\text{H}} d + \beta_{\text{H}} d_{\text{H}}^2. \tag{2.12}$$

Since the RBE is defined as the ratio of the control or reference low LET dose to the high LET test radiation dose for the same bio-effect, i.e. $\text{RBE} = d_{\text{L}}/d_{\text{H}}$, then as d approaches zero dose, by neglecting β terms in equation (1.4), the RBE is given by $\alpha_{\text{H}}/\alpha_{\text{L}}$, which is referred to as RBE_{max}. Also, for high dose, by neglecting α terms in equation (1.4), the RBE (or d_L/d_H) will approach a value of $\sqrt{(\beta_H/\beta_L)}$, which is called RBE_{min}.

It follows that

$$\text{RBE}_{\text{min}}{}^2 = \beta_H/\beta_L \tag{2.13}$$

and that

$$\beta_H = \text{RBE}_{\text{min}}{}^2\,\beta_L. \tag{2.14}$$

To obtain the high LET BED_H the entire iso-effect linear quadratic equation (1.4) must be divided throughout by α_L, the low LET α parameter, as was done in the original BED derivation. This procedure ensures that the high LET BED and the low LET BED are compatible and expressed in the same units.

To simplify further, the right hand side of the iso-effect is only considered from here on and in its fractionated form. Then the high LET BED will be given by dividing $E = n_H(\alpha_H d + \beta_H d^2)$ by α_L, to obtain

$$\text{BED} = \frac{E}{\alpha_L} = D_H\left(\frac{\alpha_H}{\alpha_L} + \frac{\beta_H}{\alpha_L} d_H\right). \tag{2.15}$$

From the above equation, where D_H is used to express the total dose given by $n_H d_H$, and then using RBE_{max} (to replace both α parameters) and RBE_{min} to replace β_H by the expression in equation (1.5), then

$$\text{BED} = D_H\left(\text{RBE}_{\text{max}} + \frac{\text{RBE}_{\text{min}}{}^2\, d_H}{\left(\frac{\alpha}{\beta}\right)_L}\right). \tag{2.16}$$

It is then possible to pursue iso-effect calculations using the equality:

$$\text{BED} = D_L\left(1 + \frac{d_L}{\left(\frac{\alpha}{\beta}\right)_L}\right) = D_H\left(\text{RBE}_{\text{max}} + \frac{\text{RBE}_{\text{min}}{}^2 d_H}{\left(\frac{\alpha}{\beta}\right)_L}\right). \tag{2.17}$$

Since $D_L = n_L d_L$, and $D_H = n_H d_H$, then if two schedules with the same number of fractions are compared, it follows that

$$d_L\left(1 + \frac{d_L}{\left(\frac{\alpha}{\beta}\right)_L}\right) = d_H\left(\text{RBE}_{\text{max}} + \frac{\text{RBE}_{\text{min}} d_H}{\left(\frac{\alpha}{\beta}\right)_L}\right). \tag{2.18}$$

One can then obtain the value of d_H from the positive root of the above equation which, with $(\alpha/\beta)_L$ replaced by k, is:

$$d_{\mathrm{H}} = \frac{-k\mathrm{RBE}_{\max} + \sqrt{k^2\mathrm{RBE}_{\max}^2 + 4d_{\mathrm{L}}{}^2\mathrm{RBE}_{\max}^2 + 4d_{\mathrm{L}}k\mathrm{RBE}_{\max}^2}}{2\mathrm{RBE}_{\max}^2}. \tag{2.19}$$

The RBE will then be $d_{\mathrm{L}}/d_{\mathrm{H}}$.

As an example, if the megavoltage photon dose per fraction is 2 Gy, and the $\mathrm{RBE}_{\max}$ is 1.6 with $\mathrm{RBE}_{\min} = 1.05$, d_{H} is found to be 1.54 Gy when $\alpha/\beta = 3$ Gy. The operative RBE will be 2/1.54 = 1.3. If the photon dose is changed to 6 Gy, then the equivalent d_{H} is 5.15 Gy and the RBE reduces to 1.16.

For dose inhomogeneity it is sometimes convenient to use a factor x, where $x = 1.07$ would indicate a 7% dose increment or $x = 0.93$, a 7% dose reduction, so that

$$\mathrm{BED} = D_{\mathrm{L}}x_{\mathrm{L}}\left(1 + \frac{d_{\mathrm{L}}x_{\mathrm{L}}}{\left(\frac{\alpha}{\beta}\right)_{\mathrm{L}}}\right) = D_{\mathrm{H}}x_{\mathrm{H}}\left(\mathrm{RBE}_{\max} + \frac{\mathrm{RBE}_{\max}^2 d_{\mathrm{H}}x_{\mathrm{H}}}{\left(\frac{\alpha}{\beta}\right)_{\mathrm{L}}}\right). \tag{2.20}$$

BEDs can, in principle, be expressed as an equivalent uniform BED just as in the case of equivalent uniform dose (Niemierko 1997), but in such cases there is loss of information on the extreme values,, which can influence outcomes; also position is lost. It is arguably better to use a BED display superimposed on 3D anatomy or using representative points where possible.

2.2.1.6 Dose rate and incomplete repair

Protraction of radiation exposure is known to reduce cell kill in most instances. Delivery of large single fractions often require a long time, due to setting up several treatment fields, with gantry rotations (which are slower because of the heavier particle therapy gantries) and treatment alignment checks. This can take as long as two hours or more in some instances, which allows considerable repair of sub-lethal damage, which has a half time of repair of around 10–15 min, with additional longer values in some bio-systems, as discussed below.

For explanatory purposes, dose rate effects are firstly expressed for low LET conditions, for dose rate R, treatment duration T, and where μ is the sub-lethal damage repair rate coefficient (linked only with the dose rate sensitive β parameter). The standard LQ equation becomes $E = \alpha RT + \beta R^2T^2 f(\mu T)$. In each case RT can be replaced by either d for a single fraction or nd for multiple fractions. Again, by dividing throughout by α, then

$$\mathrm{BED} = RT\left(1 + \frac{Rf(\mu T)}{\left(\frac{\alpha}{\beta}\right)_{\mathrm{L}}}\right). \tag{2.21}$$

The $f(\mu T)$ term used is fully expressed as

$$(2/\mu T)(1 - (1 - \exp[-\mu T])/(\mu T)), \tag{2.22}$$

which can be simplified to $2(\mu T)$ for $T > 10$ h.

For biphasic repair where there are two operative repair systems, each with different time constants μ_1 and μ_2, respectively, in proportions p and $(1 - p)$, each with incomplete repair functions f_1 and f_2, the BED equation becomes

$$\mathrm{BED} = pdRT\left(1 + \left(RT \Big/ \left(\frac{\alpha}{\beta}\right) f_1\right)\right) + (1 - p)RT\left(1 + \left(RT \Big/ \left(\frac{\alpha}{\beta}\right) f_2\right)\right). \quad (2.23)$$

The above expression is normally used for radionuclide treatments, but for external beams where the treatment is given as a dose per fraction d, each given in a fraction time t, but to an overall duration of T, then to include repair between fractions and the dose rate effect during each individual fraction, the BED can be approximated as

$$\mathrm{BED} = nd\left(1 + \left(\frac{nd}{k} - \frac{d}{k}\right) f(\mu T) + \frac{d}{k}\mathrm{f}(\mu t)\right). \quad (2.24)$$

This equation is applicable for large single fractions of dose D (= nd), when the sub-fractions d refer to exposures to different iso-centres. An additional repopulation correction factor may be required for the separate case of multiple fractionated treatments given over a long duration.

This approach can be extended to biphasic repair as:

$$\begin{aligned}\mathrm{BED} = xnd&\left(1 + \left(\frac{nd}{k} - \frac{d}{k}\right) f(\mu_1 T) + \frac{d}{k} f(\mu_1 t)\right) + (1 - x)nd \\ &\times \left(1 + \left(\frac{nd}{k} - \frac{d}{k}\right) f(\mu_2 T) + \frac{d}{k} f(\mu_2 t)\right).\end{aligned} \quad (2.25)$$

For high LET the same equations can be used with the $\mathrm{RBE}_{\min}$ terms linked to the $f(\mu T)$ or f_1 and f_2 functions, with $\mathrm{RBE}_{\max}$ replacing 1 in the relative effect part. So, for a single repair process,

$$\mathrm{BED} = nd(RBE_{\max} + (nd/k - d/k) f(\mu T)\mathrm{RBE}_{\min}^2 + d/kf(\mu t)\mathrm{RBE}_{\min}^2) \quad (2.26)$$

and for the biphasic repair processes,

$$\begin{aligned}\mathrm{BED} &= xnd\left(\mathrm{RBE}_{\max} + \left(\frac{nd}{k} - \frac{d}{k}\right) f(\mu_1 T)\mathrm{RBE}_{\min}^2 + \frac{d}{k} f(\mu_1 t)\mathrm{RBE}_{\min}^2\right) \\ &+ (1 - x)nd\left(\mathrm{RBE}_{\max} + \left(\frac{nd}{k} - \frac{d}{k}\right) f(\mu_2 T)\mathrm{RBE}_{\min}^2 + \frac{d}{k} f(\mu_2 t)\mathrm{RBE}_{\min}^2\right).\end{aligned} \quad (2.27)$$

From the data analysed by Pop *et al* (2000) and used by Hopewell *et al* (2012), the value of p is 0.495 with μ_1 as 3.647 h^{-1} and μ_2 as 0.321 h^{-1} (these represent exponential half-times of repair of 0.19 and 2.16 h, respectively), with a spinal cord α/β of 2.47 Gy. For short durations of time of up to 2.5 h, these simplified equations give results 3% above the minimum BEDs and around 4% below the average BEDs estimated previously using the more complex formulations representing incomplete repair of very closely spaced fractions, and continuous repair during the radiation exposure, as in Canney and Millar (1997).

To obtain iso-effective doses to a single fraction of megavoltage photons or γ-rays, which have times longer than a few minutes, the durations of treatment should be used. A single instantaneous dose per fraction of 12.5 Gy (γ-rays) delivered instantaneously has a BED of around 75.76 $Gy_{[2.47]}$, but when the time is extended to 0.5 h this value becomes 61.66. The iso-effective dose is 10.71 Gy for a proton beam with RBE_{max} = 1.75 and RBE_{min} = 1.09 (note that this would be 11.36 Gy if a 'standard' 1.1 RBE correction for the 12.5 Gy dose were used). For further extensions of time, iso-effective doses may be individually solved for this BED value, as shown in figure 2.2. It should be noted that the total doses required to maintain an iso-effect for α/β = 2.47 Gy (as might be the case in very slow growing tumours) increases with treatment time to match the required BED. If this is not done then tumour control could be reduced, although the fall in BED should be protective on normal tissues. The doses should be recalculated using more specific tumour α/β ratio values. Where normal tissue toxicity is of great concern, an elective decision to deliver a treatment fraction or individual fractions over a slightly longer time may be appropriate.

Readers may wish to consider using BEDs from their own dose protocols, which could be a BED given say over a designated time, such as a half or one hour using photons and then find the solution for the equivalent dose of protons at relevant LET and RBE values using equations given above, and later in chapters 7 and 8. In principle, these refinements could be added to treatment planning systems. The textbook by Wigg (2008) contains comprehensive summaries of dose rate radiobiology and practical applications, as does the shorter review by Dale and Jones (1998). Suffice it to say that tumour repair half-time rates may be shorter than in some normal tissues, but with some reduced repair capacity if there is an absence of a repair pathway, especially the recombination repair mechanism. Further work needs to be done on this important topic.

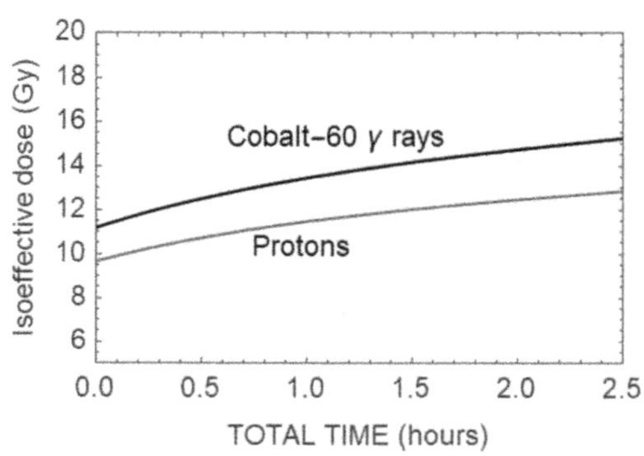

Figure 2.2. Plot of iso-effective doses and total treatment time to match the same effect (or BED) as a single dose of 12.5 Gy given by megavoltage photons or cobalt γ-rays given in 0.5 h. Note that this applies only in the case of the specified parameters used for a variable RBE.

2.2.1.7 Closely spaced fractions

Unlike the dose rate effect where $f(\mu T)$ terms cause reduced effects with increasing time, closely spaced fractions incur higher effects by imposing a multiplicative factor $(1 + h)$ to replace the $f(\mu T)$ term; readers should note that sometimes a term h is used where $h > 1$, which can cause confusion.

$$\text{The standard BED is then BED} = nd\left(1 + \frac{d(1 + h)}{\frac{\alpha}{\beta}}\right). \tag{2.28}$$

To find h values based on the mean inter-fraction interval (f), formulations and tables are available in Dale (1985) and Jones *et al* (2001). As with the dose rate effect, the overall influence of a higher LET causing larger increments in α will diminish the magnitude of closer fraction spacing to some extent, but this aspect needs careful assessment since breakdown of normal tissue tolerances have occurred with closely spaced fractions in the past, in late reacting normal tissues such as the spinal cord.

The high LET BED equation for incomplete repair becomes

$$\text{BED} = nd\left(\text{RBE}_{\text{max}} + \frac{d(1 + h)\text{RBE}^2_{\text{min}}}{\frac{\alpha}{\beta}}\right). \tag{2.29}$$

2.2.1.8 Hypoxia

For hypoxia studies, the specific oxygen enhancement ratio (OER) for each of the α_L and β_L terms can be incorporated as OER_α and OER_β, as multipliers of the dose, so that

$$\text{BED} = nd\left(\text{OER}_\alpha + \frac{d\ \text{OER}_\beta}{\frac{\alpha}{\beta}}\right). \tag{2.30}$$

The same replacements can be used within the high LET BED equation, as

$$\text{BED} = nd\left(\text{RBE}_{\text{max}}\text{OER}_\alpha + \frac{d\ \text{OER}_\beta\text{RBE}^2_{\text{min}}}{\frac{\alpha}{\beta}}\right). \tag{2.31}$$

Care must be taken to use the appropriate OER, which will itself diminish with LET. This aspect has been explored by several authors: Bopp *et al* (2016), Furusawa *et al* (2000), Wenzl and Wilkens (2011).

2.2.1.9 Very low dose

Deviations from the LQ model in fast growing cells with a high G2 cell cycle phase content exist, and have been described as the phenomenon of 'low dose hypersensitivity', thought to be related to induced repair; in other words induction of repair enzyme systems with increasing dose. These have been modelled empirically by Joiner, as in Marples *et al* (2004), but more mechanistically using the Michaelis–Menten enzyme kinetics approach and induction of additional enzyme capacity in Jones and Morgan (2007). This is relevant to high LET radiations as the extreme initial slope represents a state of enhanced radiosensitivity represented by an increase in the α coefficient, and so it represents the equivalent of a high LET situation where there is an increased yield of non-repairable damage, which does not exhibit a slightly higher surviving fraction at doses up to 1.5 Gy before resuming the normal LQ survival curve shape. For intermediate LET values one can assume intermediate patterns of this phenomenon. In most practical applications the doses will exceed the dose range (0–1.5 Gy at the most) in which this phenomenon operates.

These formulations can be transformed to higher LET situations by replacing α with either α_H or $\alpha_L R_{max}$, and β by either β_H or $\beta_L R^2_{min}$. The low LET formulations are:

$$\mathrm{SF[overall]} = f\,\mathrm{e}^{\frac{\alpha d}{\rho_{res}} - \frac{\beta d^2}{\rho_{res}}} + (1 - f)\mathrm{e}^{-\alpha d - \beta d^2} \quad (2.32)$$

where

$$\rho_{res} = x + \rho(1 - x) \quad (2.33)$$

and

$$\frac{v}{V_{max}} = \frac{\alpha d + \beta d^2}{K_M + \alpha d + \beta d^2} = \rho \quad (2.34)$$

where f is the fraction of cells in G2 phase, and x is the fraction of DNA repair that is active at zero dose. The saturation of repair rate (ρ) is described by v/V_{max} where v is the rate at dose d.

The change in effective K_M from a minimum to maximum value is described by

$$K_{M\mathrm{eff}} = K_{max} + \mathrm{e}^{-jd}(K_{M\ min} - K_{M\ max}). \quad (2.35)$$

Values of $f = 20\%$, $K_{init} = 10$, $K_{min} = 0.01$ and $j = 6$ were used to fit data in the above source reference.

The more pragmatic equation developed by Joiner *et al* (2001), shows the effective change in α with dose as

$$\mathrm{SF} = \exp\left[-\alpha_r d\left(1 + \left(\frac{\alpha_s}{\alpha_r} - 1\right)\right)\exp\left(-\frac{d}{d_C}\right) - \beta d^2\right]. \quad (2.36)$$

This is of considerable interest since as α_r (the 'normal' LQ parameter) becomes larger and approaches α_s (the more sensitive parameter, which describes the initial steep slope of the phenomenon), then the plotted function will shift towards the steep initial slope (as shown in figure 2.3). This may have relevance to high LET radiations since the α_s probably reflects the high LET condition (or the α_U, the ultimate value of α at high LET discussed in chapters 7 and 8, since there is effectively no repair at such low dose, just as there is reduced or absent repair at high dose with high LET radiations depending on the LET value). Consequently, α_r can be replaced by α_H (or

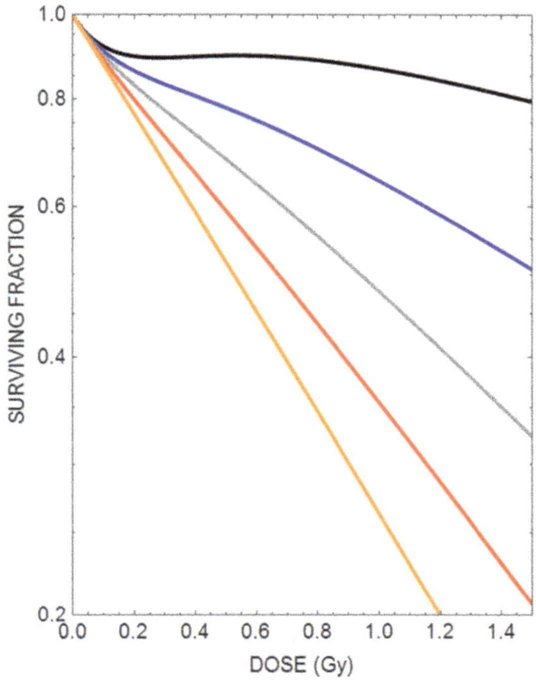

Figure 2.3. Theoretical plot of the effect of increasing the radiosensitivity (assumed here due to LET), causing progressive loss of low dose hyper-radiosensitivity. The black curve has $\alpha = 0.1\ \text{Gy}^{-1}$, blue $\alpha = 0.4\ \text{Gy}^{-1}$, grey $\alpha = 0.7\ \text{Gy}^{-1}$, and red $= \alpha = 1.0\ \text{Gy}^{-1}$. The orange line represents α_U, the maximum possible radiosensitivity for the cell line (here $\alpha_U = 1.3\ \text{Gy}^{-1}$).

by $\alpha_r \text{RBE}_{max}$) up to a limit of α_U, at any high LET. So, at the LET-RBE turnover positions representing maximum efficiency

$$\text{SF} = \exp\left[-\alpha_H d\left(1+\left(\frac{\alpha_U}{\alpha_H} - 1\right)\right)\exp\left(-\frac{d}{d_C}\right) - \beta_H d^2\right]. \tag{2.37}$$

2.2.1.10 Higher doses per fraction

There continues to be discussion about deviations from the LQ model at higher doses, even if this may be an artefact due to experimental conditions. In different cell systems and growth conditions, deviations have been observed at doses above the 6–10 Gy range. This effect can occur at least partly because of the statistical issues that arise for lower surviving fractions in colony forming assays where the number of colonies growing is smaller. Nevertheless, there are also some theoretical reasons why some apparent straightening of the cell survival curve might occur with increasing dose.

Two approaches are presented here. The first assumes that second or third lethal chromosomal breaks on the same chromosome result in 'wasted dose', being incapable of causing additional lethality with resulting high dose inefficiency of the survival curve. The second is a pragmatic model, which assumes a final slope

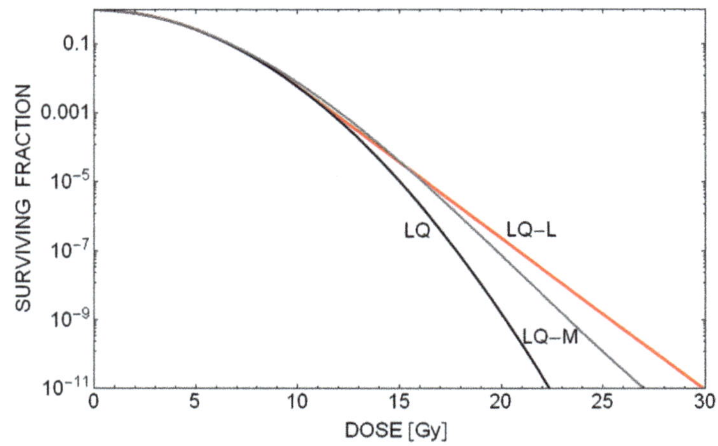

Figure 2.4. Three theoretical survival curves taken to high doses capable of causing tumour sterilisation. The curves appear to deviate around 10 Gy. The standard LQ model shows progressive curvature, the LQ-M (chromosomal) model 1 gradually straightens and the LQ-L (or LQ-linear) model 2 follows a constant slope after a dose of 10 Gy has been achieved.

beyond a fixed dose. Alternative approaches include that of Park *et al* 2008 who suggested methods for a gradual transition of the dose response from the LQ to increasing linearity at higher doses. It is claimed that such approaches may be better suited for assessing high-dose radiotherapy, but others (e.g. Fowler 2008, 2010) suggest that caution needs to be exercised in their application.

To return to the two models presented here, they are shown graphically with a comparison to the standard LQ model in figure 2.4.

The yield of lethal injury is conventionally represented by the number of lethal chromosome breaks per cell, broadly represented by $\alpha d + \beta d^2$, as initially determined by Douglas Lea in 1946. Since some of these breaks will be second or third breaks on the same chromosomes (this being likely since some chromosomes are larger than others), such an effect would require modification of the yield of 'effective' lethal chromosomal events, defined as lethal breaks restricted to only one chromosome (second, third, etc breaks on the same chromosome representing no further lethality risk). This makes a difference when Poisson statistics are considered, and such an interpretation is given in table 2.2.

The model provides a relatively simple formulation but requires radiosensitivity and chromosomal number input. Straightening of the LQ curve occurs at reasonable doses in the 6–10 Gy range depending on the radiosensitivities and scaling used. This model may be appropriate for normal tissue with a normal chromosome content of M = 46. Rapidly dividing cells have the added complexity of chromatids, and tumour cells may possess greater than 46 chromosomes per cell. Another complexity is that chromosomal volumes are variable, so some chromosomes will be more susceptible than others to random radiation damage. The chances of lethal breaks and multiple breaks on the smallest chromosomes can probably be ignored so one might compromise on there being an effective smaller input number M to the model.

Table 2.2. Derivation of chromosomal number considerations in apparent straightening of the cell survival curve with increasing dose.

- The nuclear chromosome number is M.
- Total number of lethal chromosome breaks per cell, $E = \alpha d + \beta d^2$.
- Expected average number of lethal breaks per chromosome at dose $d = (\alpha d + \beta d^2)/M$.
- Probability of *NO* breaks on an individual chromosome is $\exp[-(\alpha d + \beta d^2)/M]$.
- Probability of any number of lethal breaks on an individual chromosome is $1 - \exp[-(\alpha d + \beta d^2)/M]$.
- The average number of lethally damaged chromosomes per cell becomes $M(1 - \exp[-(\alpha d + \beta d^2)/M]) = E_M$.
- So the surviving fraction becomes $\mathrm{SF} = \exp[-E_M]$.

A number of around 36–40 is suggested. Some unpublished work has shown that a greater number of cell survival curves were better fitted by this model than the standard LQ model, but the difference was not statistically significant. However, the model is attractive for normal tissue modelling purposes and may be biologically more realistic than the assumption of an arbitrary dose beyond which a straight line is assumed, which is to be considered next.

Since the chromosomes are of variable volume, random lethal break distributions will favour the larger ones, so there is likely to be an 'effective value' that is less than the 46 chromosomes of a cell. This is not known at the present time but is potentially obtainable by experiment. A disadvantage of this model is that it is not possible to use the BED concept with accuracy; this is because series expansion of the exponential term does not converge well on the pure function, and there is increasing dependence on knowledge of the α and β terms rather than α/β. Consequently, graphical methods are necessary to determine the reduction of effect or the equivalent dose to produce the same effect, and then apply BED equations. The method in its present form may be a useful research tool, since by fitting appropriate data sets the effective value of M can be obtained.

Jones and Dale (2015) presented an alternative simple approach, inclusive of a BED approach, which assumes that the transition from pure LQ cell kill to purely logarithmic (i.e. straight line) cell kill is abrupt rather than gradual and occurs at a specific fractional dose (c) after which the same slope is maintained. In terms of the number (L) of lethal lesions created.

At low fraction doses (less than c) equation (1) holds, i.e. $E = \alpha d + \beta d^2$ up to a limit of $E = \alpha c + \beta c^2$, but at a dose of c the gradient is given by the first differential coefficient of the linear quadratic equation ($\alpha + 2\beta c$), so that for doses that exceed c (i.e. $d - c$):

$$E = \alpha c + \beta c^2 + (\alpha + 2\beta c)(d - c) \tag{2.38}$$

which provides a constant slope for $d > c$.

Following exactly the same procedures as were applied in the basic BED derivation (see equations (1)–(5)), the BED for N fraction treatments involving fraction doses greater than c is given by:

$$\mathrm{BED}_N = N\left(c + \frac{c^2}{(\alpha/\beta)} + \left(1 + \frac{2c}{(\alpha/\beta)}\right)(d - c)\right) \tag{2.39}$$

which simplifies to:

$$\mathrm{BED}_N = Nd\left(1 + \frac{c}{(\alpha/\beta)}\left(2 - \frac{c}{d}\right)\right) \tag{2.40}$$

For purposes of illustration, we assume that the crossover from LQ cell kill to linear (logarithmic) cell kill occurs at 10 Gy, i.e. $c = 10$. Then, again following the steps used earlier, for a single fraction treatment to be iso-effective for a conventional 30 × 2 Gy treatment (BED = 72 Gy_{10}), then,

$$1 \times d \times \left(1 + \frac{10}{10}\left(2 - \frac{10}{d}\right)\right) = 72$$

i.e. the single fraction iso-effective tumour dose (d) is 27.3 Gy (cf 22.3 Gy derived above for the 'pure-LQ' case).

In the case of a late reacting normal tissue with $\alpha/\beta = 3$ Gy (for 60 Gy in 30 fractions, BED = 100 Gy_3), we then solve

$$1 \times d \times \left(1 + \frac{10}{3}\left(2 - \frac{10}{d}\right)\right) = 100$$

which provides a single fraction dose of 17.4 Gy (cf 15.9 Gy derived for the 'pure-LQ' case).

The effect of changing the dose at which the transition occurs is shown in figure 2.5.

High LET formulations for each method

1. Chromosomal M model: use the BED formulation containing RBE_{max} and RBE_{max} instead of the standard BEDs:

   $E_M = M(1 - \exp[-(\alpha_H d + \beta_H d^2)/M])$ or $E_M = M(1 - \exp[-(\alpha_L R_{max} D + \beta_L \mathrm{RBE}^2_{min} d^2)/M])$

   where M is the 'effective number of chromosomes'.
2. For the straightening of the LQ curve after a dose c, it can be shown by similar rearrangements, as in the standard model case, that

$$\mathrm{BED}_N = Nd\left(\mathrm{RBE}_{max} + \frac{\mathrm{RBE}^2_{min}\, c}{(\alpha/\beta)}\left(2 - \frac{c}{d}\right)\right). \tag{2.41}$$

In very high LET states, if the effect curvature is reduced or absent, then the models described may only need to be applied for the reference radiation in RBE studies.

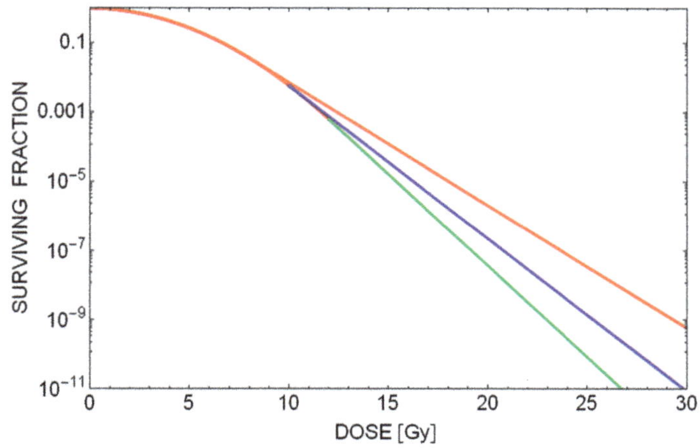

Figure 2.5. Changing the transition points at 8 Gy (red), 10 Gy (blue) and 12 Gy (green).

It is important to understand that standard BED equations (those assuming that the LQ model holds for indefinitely large fraction doses) will always under-estimate the iso-effective doses required for treatments involving high fraction doses. However, caution must be advocated in applying this alternative and un-tested BED approach, but it might be useful for those who wish to analyse their data and attempt to isolate what is the best dose for c in particular situations such as normal tissue type or tumour class.

In practice, because the dose–response curves 'straighten out' (compared with the increasing curvature predicted by the LQ model), the iso-effect doses calculated using the above methodology provide lower tolerance or iso-effective dose estimates compared with the standard LQ approach. To that extent the standard BED calculations tend to be 'fail-safe'; both the required tumour and normal tissue doses will be under-estimated, but the degree of under-estimation is probably greater in the latter case. This is because the late reacting normal tissue dose–response curve has greater curvature (and lower α/β ratio) than has the tumour. In the case of slow growing tumours with α/β ratios that are close to that of the late reacting normal tissue, there should be less difference. However the LQ model will continue to under-estimate the effect and will be fail-safe with respect to normal tissue late reacting effects. This must be understood by the physicians and physicists involved in the treatment planning process. These methods may of course be useful in obtaining correlations with clinical data sets.

The inclusion of normal tissue and tumour repopulation effects, and furthermore detailed considerations on fractionation, and the increasingly important topic of retreatment of recurrent tumours, are discussed further in chapters 5 and 10.

2.3 The α/β ratio and its choice for modelling particle therapies

Models can only be sufficiently informative for clinical decision making if their input parameters are reasonably accurate and if the model is used in a 'fail-safe' way, sometimes varying these parameters to consider worse case scenarios. The radiobiology literature contains many examples of such parameters, especially the key α/β ratio of important normal tissues and certain tumour types, although in the latter case with considerable variation. Unfortunately, good predictive assays are not available, although it is clear that correlations between concepts like repair capacity will correlate with radiosensitivity and that α and β may reflect some of the limiting repair processes such as NHEJ and RR, respectively. Much work is needed to establish firm links between molecular based assays and the modelling parameters.

2.3.1 The α/β ratio

The α/β ratio is the ratio of the two radiosensitivity parameters in the LQ model of the radiation effect and is expressed in units of Gy. The ratio is

- a relatively robust parameter, especially in the case of late reacting normal tissues with little change between species,
- inversely related to the fraction sensitivity during radiotherapy,
- well characterised for certain classes of tissue and tumours in experimental cell lines, animals and humans,
- inversely proportional to cell turnover.

The α/β ratio probably reflects many cellular processes, such as DNA repair and changes in radiosensitivity due to cell cycle position, and in the case of tumours the extent of hypoxia, although a single α/β value can only represent the average contributions of these processes during therapy. It is an especially useful parameter within BED equations.

The function of the α/β ratio is to act as a coefficient that controls the fractionation sensitivity. When fraction size is changed, the resulting total dose needs to be modified in order to provide a constant BED (and the same bio-effect). It is important to understand that different tissues and tumours can possess quite different α/β ratios, so that they respond differently to different changes in dose fractionation. The use of α/β ratios in hadrontherapy is more complex, as discussed below.

2.3.2 Applications of BED equations

There are many uses, including the calculation of

- iso-effect (or equi-effect) doses,
- compensatory doses after unintended under- or over-dosage,
- modified dose fractionation after unintended treatment interruptions.

BED equations can be adapted for the purposes of charged particle therapy using protons and ions, as well as for drug (chemotherapy) related effects and for the influence of changes in radiation tolerance due to factors such as age/surgery by allocating a BED equivalent for these factors, as shown below.

BED equations are derived from the basic linear quadratic model of radiation effect by dividing the number of lethal events per cell by the α parameter. Although the BED equation consists of a simple mathematical (quadratic) form, clinical BED calculations require considerable care, especially in the choice of α/β ratio used; this applies to both conventional x-ray based megavoltage radiations and charged particle therapy using hadrons.

There are some important rules concerning BED equations.

1. BEDs can be added for different components of an overall treatment given by different types of radiation (e.g. for a treatment combining x-rays and electrons, the overall BED = electron BED + x-ray BED, or for a treatment comprising x-rays and an ion beam boost, then the overall BED = x-ray BED + ion beam BED).
2. Additional risk factors such as chemotherapy, age or other medical conditions can be included as overall BED = radiation BED + risk factor BED. The latter is determined from clinical studies that contain multiple radiation dose levels with and without the risk factor, where an iso-effect can be identified, and allowing simultaneous equations to be used as shown by Jones *et al* (2006a).
3. Calculations that involve a specified bio-effect in a given tissue or tumour type should be done only using the same α/β ratio; it follows that different α/β ratios cannot be used to describe the same bio-effect.
4. It is assumed that the α/β ratio reflects an average value during treatment. During a course of radiotherapy lasting 4–8 weeks, the ratio is probably stable in the case of slow growing tumours and slow cell turnover tissues. However, in the case of tissues or fast growing tumours and normal tissues that exhibit real or apparent acceleration of their proliferation rates during treatment, the α/β ratio may itself increase; this has been shown to be the case for acute reacting normal epithelial tissues, where α/β increases from 5 to 35 Gy from the first few weeks to the sixth week in human and animal skin (Hopewell *et al* 2003).

2.3.3 Special considerations for particle therapy

A reader should be cautious about some published high-LET BED values, since they have incorporated a constant RBE value. The literature can be confusing, as this work can still be regarded as being in a developmental stage despite many of the underlying principles being understood around 40 years ago, but before the more widespread utilisation of the LQ model and BED concept. Subsequent improvements have now produced a more coherent framework for iso-effect applications. Two suggested methods for BED iso-effect calculations in modern hadrontherapy are provided below; these use both low and high LET α/β ratios in specific situations, but principally a low LET α/β in association with RBE conversion factors. It must be stressed that dose in these calculations refers to the physical dose and not the cobalt equivalent or any other form of RBE weighted dose, unless otherwise stated.

The relationship between $(\alpha/\beta)_L$ and $(\alpha/\beta)_H$ is linked in the following way by the maximum and minimum RBE parameters:

$$\frac{\mathrm{RBE}_{\mathrm{max}}}{\mathrm{RBE}_{\mathrm{max}}^{2}} = \frac{\left(\dfrac{\alpha}{\beta}\right)_{\mathrm{H}}}{\left(\dfrac{\alpha}{\beta}\right)_{\mathrm{L}}}. \tag{2.42}$$

It also follows that

$$\left(\frac{\alpha}{\beta}\right)_{\mathrm{H}} = \frac{\mathrm{RBE}_{\mathrm{max}}}{\mathrm{RBE}_{\mathrm{max}}^{2}}\left(\frac{\alpha}{\beta}\right)_{\mathrm{L}}. \tag{2.43}$$

This provides a general method for estimating the α/β for high LET radiations where these are not known. This equation can be further simplified by use of a combined RBE converting factor (R_C) as

$$\left(\frac{\alpha}{\beta}\right)_{\mathrm{H}} = R_C\left(\frac{\alpha}{\beta}\right)_{\mathrm{L}} \tag{2.44}$$

where

$$R_C = \frac{\mathrm{RBE}_{\mathrm{max}}}{\mathrm{RBE}_{\mathrm{max}}^{2}}. \tag{2.45}$$

Consequently a single number, R_C, can be used to convert low LET α/β to the high LET state.

The increase in RBE, when the dose per fraction is reduced, is mainly due to the increase in the repair of sub-lethal damage that occurs in the low LET radiation case. Then, in principle, $\mathrm{RBE}_{\mathrm{max}}$ and $\mathrm{RBE}_{\mathrm{min}}$ may also be governed in part by the low LET α/β ratio, such that a greater change in RBE with dose per fraction occurs in tissues with lower low LET α/β ratios (due to their ability to repair low LET radiation sub-lethal damage more effectively than tissues with higher α/β values). Recent analysis of fast neutron data sets in the UK has shown (for cells and tissues that are not seriously repair deficient) that the $\mathrm{RBE}_{\mathrm{max}}$ and $\mathrm{RBE}_{\mathrm{min}}$ are, respectively, inversely and directly related to the low LET α/β ratio. $\mathrm{RBE}_{\mathrm{max}}$ appears to be a reciprocal function of the low LET α/β ratio, but $\mathrm{RBE}_{\mathrm{min}}$ is linearly related to the square root of the low LET α/β ratio. In other words, RBE is not only a combination of change in LET and dose (physical), but is also influenced by changes in cell cycling/repair (biological), as reflected in the α/β ratio.

It follows from the definitions of $\mathrm{RBE}_{\mathrm{max}}$ and $\mathrm{RBE}_{\mathrm{min}}$ and equation (2.39) that

$$\mathrm{RBE}_{\mathrm{max}} = \mathrm{RBE}_{\mathrm{min}}^{2}\frac{\alpha_{\mathrm{H}}}{\beta_{\mathrm{H}}}\left(\frac{1}{(\alpha/\beta)_{\mathrm{L}}}\right) \tag{2.46}$$

and

$$\mathrm{RBE}_{\mathrm{min}} = \sqrt{\frac{\mathrm{RBE}_{\mathrm{max}}(\alpha/\beta)_{\mathrm{L}}}{(\alpha/\beta)_{\mathrm{H}}}}. \tag{2.47}$$

It can clearly be seen that $\mathrm{RBE}_{\mathrm{max}}$ is inversely proportional to the low LET α/β and that $\mathrm{RBE}_{\mathrm{min}}$ is directly proportional to its square root (all other terms being kept constant).

In biological data, the general relationships between RBE_{max}, RBE_{min} and $(\alpha/\beta)_L$ take the following form, where *A* and *C* represent the minimum values of RBE_{max} and RBE_{min}, respectively, and *B* and *K* are coefficients

$$RBE_{max} = A + \frac{B}{\left(\frac{\alpha}{\beta}\right)_L} \tag{2.48}$$

$$RBE_{min} = C + K\sqrt{\left(\frac{\alpha}{\beta}\right)_L}. \tag{2.49}$$

Such relationships can only apply for specific LET determined situations and the above equations were derived from fast neutron data (Jones *et al* 2011), consequently from an LET spectrum, although the RBE values obtained are close to those found towards the end of the range of the SOBPs (Britten *et al* 2013), not surprisingly perhaps, since fast neutrons ionise mainly by formation of recoil protons. For many years neutron RBEs were used to provide RBEs for C ion therapy in Japan (Kanai *et al* 1997). So, the values of *A*, *B*, *C* and *K* are available for one fast neutron data set but remain to be determined for protons and heavier ions. Studies of how the basic physics-related RBE changes (*A* and *C*) with LET, the further parameters (*B* and *K*) and how they change between cells and tissues might offer a simple pragmatic approach to data fitting, but the constants would apply only to specific LET values. However, it would be preferable to derive RBE limits directly from LET, as will be shown in later chapters.

The local effect model (Elsässer *et al* 2008), developed and used in Germany, predicts RBE from LET and for C ions uses the low LET α/β ratio along with a conversion factor based on beam microdosimetry (LET) and mean cellular nuclear volume. It essentially provides an RBE which is related to the low LET α/β ratio, which varies with dose per fraction within its accepted range based on *in vitro* cell survival data and has been applied successfully in clinical treatments of carbon ions at Darmstadt and Heidelberg using 3 Gy per fraction equivalent dose at the tumour target. The RBE is said to be under-estimated by around 10–20%, which ensures that the tumour cell kill will be higher than expected from the stated equivalent dose. It remains to be seen if the model will hold when larger fraction sizes beyond those used to obtain *in vitro* data are used. In the MKM model, developed in the USA and now used in Japan (Hawkins, 2003, 2009, Inaniwa *et al* 2013), the predicted RBE also has a similar inverse dependency on the low LET α/β, and could be similarly susceptible in hypo-fractionated conditions. To verify models over a wider range of fractionation may require animal models rather than *in vitro* experiments since the amount of cell killing required to produce late tissue effects exceeds the *in vitro* range by several orders of magnitude. The alternative is to analyse human clinical data in greater detail with respect to LET, particle dose (BED) and tissue tolerances.

It is important to note that the α/β ratio increases nonlinearly with LET. This can cause confusion in modelling. Any attempt to convert between low and high LET

α/β should be done for the two components of the LQ model separately, that is for RBE_{max} (for α) and RBE_{min} (for β)

Alternative approach for iso-effect calculations in the case of two high LET schedules

Rather than convert dose information between low and high LET conditions, there will be situations where only the high LET radiosensitivity parameters are known. Then one can use the following approach without reference to low LET. Thus for two iso-effective schedules of high LET we have for N_1 factions and N_2 fractions of dose d_1 and d_2

$$N_{1H}(\alpha_H d_{1H} + \beta_H d_1^2{}_H) = N_{2H}(\alpha_H d_{2H} + \beta_H d_2^2{}_H). \tag{2.50}$$

It is then permissible to divide throughout by α_H and obtain

$$D_{1H}\left(1 + \frac{d_{1H}}{\left(\frac{\alpha}{\beta}\right)_H}\right) = D_{2H}\left(1 + \frac{d_{1H}}{\left(\frac{\alpha}{\beta}\right)_H}\right). \tag{2.51}$$

Here the α/β is that of the high LET parameters (these are less well established in the literature at the present time) but in this case the two RBE parameters RBE_{max} and RBE_{min} are no longer necessary. It would also be possible to use the single R_C conversion factor in the situation of comparing two identical high LET radiations, so that equation (2.15) would become

$$D_{1H}\left(1 + \frac{d_{1H}}{R_C\left(\frac{\alpha}{\beta}\right)_L}\right) = D_{2H}\left(1 + \frac{d_{2H}}{R_C\left(\frac{\alpha}{\beta}\right)_L}\right) \tag{2.52}$$

which could also be used in this form along with the time factor corrections, as given below.

For unintended treatment interruptions, the most appropriate time and repopulation correction factors can be added for the case of tumour iso-effects. This is not necessary for late reacting tissues. The equations are then:

$$D_{1H}\left(1 + \frac{d_{1H}}{\left(\frac{\alpha}{\beta}\right)_H}\right) - K_H T_{1H} = D_{2H}\left(1 + \frac{d_{2H}}{\left(\frac{\alpha}{\beta}\right)_H}\right) - K_H T_{2H} \tag{2.53}$$

where T_1 and T_2 are the respective overall treatment times for schedule 1 and 2, and K_H is defined as

$$K_H = \frac{0.693}{\alpha_H \cdot \omega}. \tag{2.54}$$

Compared with the K values for low LET (K_L), there is little knowledge of K_H values in the literature. K_H can however be found from K_L by dividing by RBE_{max} (which is α_H/α_L), that is

$$K_H = \frac{K_L}{RBE_{max}}. \tag{2.55}$$

For $(\alpha/\beta)_H$ there is little in the way of large scale clinical data for protons and heavier ions, although it is well established that α values increase with LET to a far greater extent than β values, which results in $(\alpha/\beta)_H$ being much larger than $(\alpha/\beta)_L$, as shown in large *in vitro* data sets. It is also possible to infer α/β values from sample clinical data. For example, the Japanese NIRS (Chiba) carbon ion phase 1/2 non-small cell lung cancer studies showed an iso-effective tumour control for 44 Gy-eq in one fraction and 60 Gy-eq in four fractions given in one week (Okada *et al* 2010); then, $(\alpha/\beta)_H$ can be found in the following equation, where an RBE of 2.5 is assumed to convert to physical dose in the middle of the SOBP for C ions at NIRS (Chiba).

$$\frac{44}{2.5}\left(1 + \frac{44}{(2.5)\left(\frac{\alpha}{\beta}\right)_H}\right) = \frac{60}{2.5}\left(1 + \frac{60}{(4 \times 2.5)\left(\frac{\alpha}{\beta}\right)_H}\right),$$

from which $(\frac{\alpha}{\beta})_H = 25.9$ Gy.

It is then possible to re-estimate this parameter from iso-effective tumour control in patients treated with either 1, 4, 9 or 18 fractions to cobalt equivalent Gy total doses of 44 (in 1 day), 60 (in 1 week), 72 (in 3 weeks) and 86 (in 6 weeks), respectively, and allowing for repopulation at an assumed equivalent rate of 0.1 Gy per day between one and three weeks and of 0.3 Gy per day at times over 28 days. These repopulation rates are in K_H units and are consequently lower than the K_L values usually seen. The total dose solutions are then 25.9, 22.2, 23.9, 15.3, 19.1 and 37.7 Gy when each schedule is compared with another. This provides a mean $(\alpha/\beta)_H$ of 24 Gy (median 23.1) with a standard error of the mean of 3.3 Gy. This would imply an R_C value of 2.4 (if the $(\alpha/\beta)_L$ is 10 Gy). Then, by using equation (39), RBE_{min} is not unreasonably around 1.3; it follows that RBE_{max} is around 4.

It can be appreciated from the previous paragraph that changes in total dose are required to compensate for altered dose per fraction even in the case of high LET radiations, which have high $(\alpha/\beta)_H$ values. It is only when high LET and very low LET radiations are compared that the fractionation sensitivity of the high LET case is considered to be almost zero in relative terms; this is reflected by the differences in their α/β ratios.

Some recommendations for future work include.

- RBE_{max} and RBE_{min} values need to be better established for high LET radiations and linked to specific LET values.
- Specific high LET repopulation dose equivalents are required for different classes of tumours.

- α/β values specific for high LET radiations can be used in some circumstances without RBE when identical high LET radiations are compared, but not for comparing low and high LET schedules, or for calculating a combined BED where low and high LET radiations are used. A library of *clinical* $(\alpha/\beta)_H$ values must be gathered for protons and ions and would need cooperative international research, although the process has commenced in Germany, at GSI, Darmstadt (Friedrich *et al* 2013) and should be extended from only *in vitro* data to *in vivo* conditions. It will take a long time to establish a comprehensive set of radiobiological parameters specific for hadrontherapy under the circumstances of using only a physical dose, the high LET parameters and no RBE factors. This is a reasonable long-term goal, which will need to also include the defining of tissue tolerances and expected tumour control rates using such an approach. In the meantime, the use of RBE to convert known tolerance and tumour prescription doses remains the norm. The only alternative is to use data from Japan where the dose per fraction has been varied and where good iso-effective dose data are available.
- Users should beware of over reliance on *in vitro* α and β values; these are acquired using ideal cell culture conditions and in only one treatment fraction, which can hardly reflect the conditions in a tumour or normal tissue with all the changes that inevitably occur during fractionated treatment.

In situations where under- or overdose has occurred, or in calculation of x-ray equivalent schedules, the most reasonable estimates of the parameters should be used; clinicians may prefer to use the safest choices, such as a high RBE_{max} in tissues that are highly fraction sensitive for x-rays, such as the central nervous system. The general advice for such calculations in the case of x-rays has been published elsewhere (Jones and Dale 2008), but the additional effect of RBE must be included if there is any attempt to use (low LET) x-ray tissue tolerances or x-ray tumour control data. The alternative approach is to use only high LET α/β ratios within BED equations where the tolerance and tumour control BED values for the high LET radiation are known. These are subtle but important differences.

The low LET α/β ratios for various tumours can be found in publications by Wyatt *et al* (2003) and Wigg (2008). It is unfortunate that, because of a worldwide reduction in radiation research applied to pure radiobiology over the past twenty years, relatively little progress has been made to determine even better parameters from clinical data sets. Despite such setbacks, this remains an important area for research.

From the clinical experience of the present author, who has implemented many forms of the LQ model in clinical situations, it is recommended that a range of α/β values should be modelled. The variation in α/β will normally be greater for tumours than for normal tissues, since the former contain greater genetic and phenotypic heterogeneity. Another issue is that α/β may vary with age and be smaller in younger individuals, consequently making them more fraction sensitive; alternatively an equivalent BED for age can be used. The responsible clinician should ideally be given a choice of a range of solutions to any clinical problem, calculated using a reasonable range of α/β values. It should also be noted that, more recently, the α/β values for human tumours are considered to be lower than those previously established from animal experiments. So, a value of 5–8 Gy may be more reasonable than 10 Gy for many human cancers. Fowler stated that the fractionation effects were far less substantial for a tumour α/β greater than 7–8 Gy when compared with normal tissues with values of between 2 and 4 Gy.

Some normal tissues probably exhibit dual response characteristics where the combined effects of severe acute reactions can make a significant contribution to the late effects; these are termed consequential late reactions. They may be the weighted product of the acute and late reacting α/β ratios and so can be greater than 3 Gy. An example is rectal damage, often quoted as 4 Gy, but in practical (and conservative) terms assumed to be 3 Gy. There probably are differences in terms of duration of radiation exposure. For the extreme example, the acute exposure of a large single fraction, we should assume the lowest α/β value since irradiation takes place in the situation of (relatively) low proliferation. However, in the case of a more protracted irradiation, the acute reacting epithelial tissue may increase its own α/β ratio with time; this is known to be the case in animal and human skin. Another important issue is that overall treatment time and dose per fraction, taken together, may influence the severity of the acute reaction sufficiently to have an impact on the late reaction. This aspect of clinical radiobiology is relatively neglected in recent times and deserves further attention.

It cannot be overemphasised that α/β ratio values in tumours need to be expressed as a function of their degree of de-differentiation, or in other words related to the complexity of genetic disorder, or chaos, within the tumour cell; in other words a choice of a larger value within the accepted range should be considered for poorly differentiated tumours and lower values for the well differentiated state. The α/β ratios of normal tissues with respect to their late reacting complications are considered to be more stable so that 2 Gy for central nervous system, and 3 Gy for all other tissues tend to be used.

Obtaining reliable ranges of K values and, where applicable, T_K values for specific tumour and normal tissue classes are also important research aims.

The following data was compiled by the present author and Dr Joe Martin as part of an introductory D Phil project (as well as figures 2.5 and 2.6). Considerable use has been made of data within Wigg (2008) page 28, table 1.3, and other sources.

Tissues and species	α/β	Upper (95%) confidence limit	Lower (95%) confidence limit
Head and neck	10.0	n/a	n/a
Prostate cancer	3.1	2.6	3.6
Prostate cancer	1.5	0.8	2.2
Prostate cancer	1.49	1.25	1.76
Melanoma	0.6	−1.1	2.5
Melanoma	0.17		

In vitro information, pooled data, page 215 of Wigg (2008).

Tumour	α	Standard deviation	β	Standard deviation	α/β
Melanoma	0.27		0.042		6.43
	0.33		0.038		8.68
	0.13		0.113		1.15
	0.27		0.047		5.74
	0.21		0.1		2.10
	0.28		0.087		3.22
	0.35		0.070		5.00
	0.53		0.078		6.79
	0.61		0.101		6.04
	0.33		0.048		6.88
Mean	0.33	0.22	0.072	0.027	4.58
Pancreatic	0.51		0.048		10.63
	0.47		0.052		9.04
Mean	0.49	0.028	0.050	0.003	9.80
Breast	0.51		0.031		16.45
	0.47		0.091		5.16
Mean	0.49	0.14	0.061	0.042	8.03
Prostate	0.44	0.27	0.037	0.028	11.89
Cervical	0.43	0.19	0.036	0.024	11.94
Colon	0.23		0.048		4.79
Astrocytoma gd.1	0.37		0.11		3.36
Astrocytoma gd.3	0.26		0.034		7.65
Glioblastoma	0.28	0.049	0.045	0.022	6.22
Glioma (paed)	0.30	0.20	0.030	0.018	10.00
Squamous cell carcinoma	0.36	0.22	0.042	0.023	8.57
Adenocarcinoma	0.45	0.32	0.039	0.033	11.54
Embryonal	0.61	0.24	0.029	0.054	21.03
Lymph/myeloma	0.56	0.25	0.085	0.102	6.59
Oat cell carcinoma	0.58	0.27	0.087	0.122	6.67
Osteosarcoma	0.38	0.19	0.063	0.038	6.03
Neuroblastoma	0.75	0.33	0.059	0.062	12.71
Medulloblastoma	0.55		0.019		28.95
Wilm's Tumour	0.55		0.023		23.91
Ewing's sarcoma	0.42	0.17	0.127	0.19	3.31
Rhabdomyosarcoma	0.25		0.029		8.62
Burkitts lymphoma	0.87		0.009		96.67
Acute myeloma	1.24	1.2	0.052	0.083	23.85
Leukemia					
ALL	0.59	0.19	0.012	0.017	49.17
AML	0.96	0.60	0.021	0.012	45.71
Chronic ML	0.09		0.007		12.86
Leukemic cell	0.65		0.065		10.00
Promyelocytic	0.59	0.056	0.028	0.039	21.07

APL	0.44		0.018		24.44
Erythroleukemia	0.53	0.36	0.031	0.042	17.10
Acute non-lymph leukaemia	1.45		0.13		11.15

In vivo human tumours		Wigg (2008) page 210	
Tissue and species	α/β	Lower (95%) confidence limit	Upper (95%) confidence limit
Oral cavity	10.0	6.5	10.3
Vocal chord	9.9		
Bladder cancer	10		
Cervix cancer	13.9		
Skin cancer	8.4		
Liposarcoma	0.4	−1.4	54
Melanoma	0.6	−1.1	2.5
Melanoma	0.17		

Values from Basic Clinical Radiobiology (Bentzen and Baumann 2002).

Tumours	α/β	Lower (95%) confidence limit	Upper (95%) confidence limit
Head and neck			
Larynx	14.5	4.9	24
Vocal chord	13		
Oropharynx	16		
Buccal mucosa	6.6	2.9	Infinite
Tonsil	7.2	3.6	Infinite
Nasopharynx	16	−11	43
Various	10.5	6.5	29
Skin	8.5	4.5	11.3

Other analytical techniques have been used for tumours such as glioblastoma where the α/β has been estimated to be 9.2 Gy (Jones and Sanghera, 2007) and there is greater acceptance that α/β will vary between tumours according to their proliferative and repair capacity; these may correlate with histological grading features, the more well differentiated the lower the ratio.

Where α/β is not known but tumour potential doubling time (and volume doubling time) information is available, one can resort to a simple inverse relationship, given by $\alpha/\beta = 48.8/T_{pot}$, as derived from the plot shown in figure 2.6. This formulation was derived from the equation $K = 0.693/(\alpha T_{eff})$, since it can be divided throughout by β and re-arranged to give $\alpha/\beta = C/T_{pot}$, where C is regarded as a replacement constant and which assumes T_{pot} is the operative T_{eff} in most

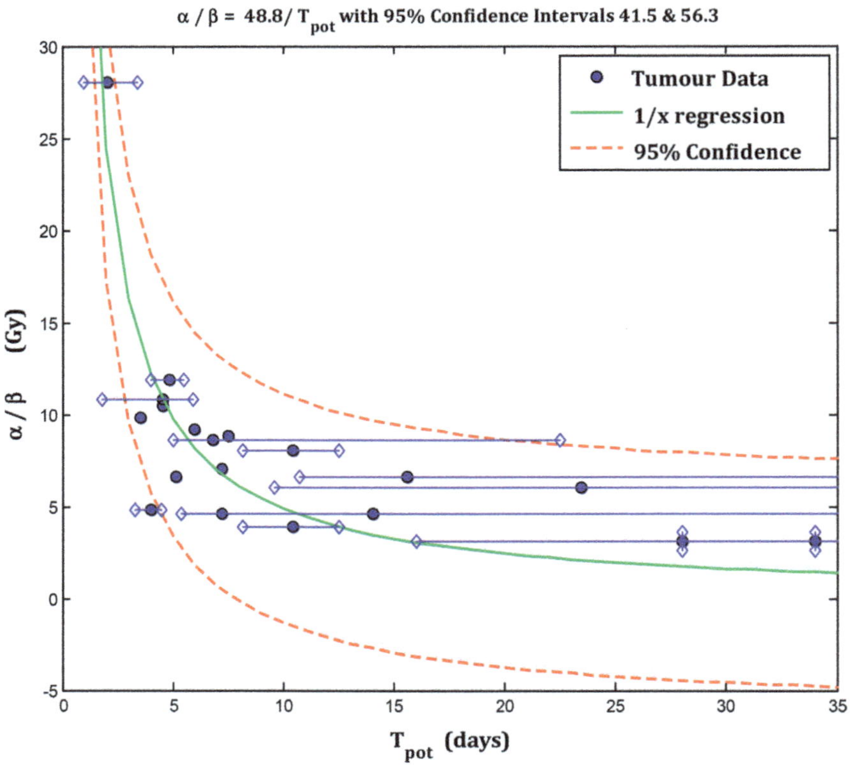

Figure 2.6. Plotted data from the literature of the cellular potential doubling time (T_{pot}) against the α/β ratio and fitted using the nonlinear least squares technique.

radiotherapy schedules. This is consistent with standard radiobiology dogma that low proliferative states have low α/β and vice versa. For measured tumour volume doubling times (T_D), one can assume a high cell loss factor of 90% for carcinomas (50% for sarcomas), which allows the estimation of the potential doubling time. The cell loss factor (ϕ) is known to be related to T_{pot} by the relationship $\varphi = (1 - \frac{T_{pot}}{T_D})\%$.

In the case of normal tissue, it is standard practice to use $\alpha/\beta = 3$ Gy for most late effects (fibrosis in lung, kidney, soft tissues, heart) but the lower value of 2 Gy in the central nervous system (spinal cord and brain). Peripheral nerves and the cauda equina (below the spinal cord, but consisting of nerves derived from it) may have α/β between 2 and 3 Gy although the more conservative 2 Gy is often used. Another important area is the optic chiasm where α/β of 2 Gy should be used, whereas the optic nerve may behave more like peripheral nerve.

Using the above equations and assumption of the cell loss factor linking T_{pot} and volume doubling times, it is possible to build up 3D plots of the relationship between volume doubling times and RBE in the case of fast neutrons (see figure 2.7). This approach may reflect what is also operative within the Bragg peak region for protons.

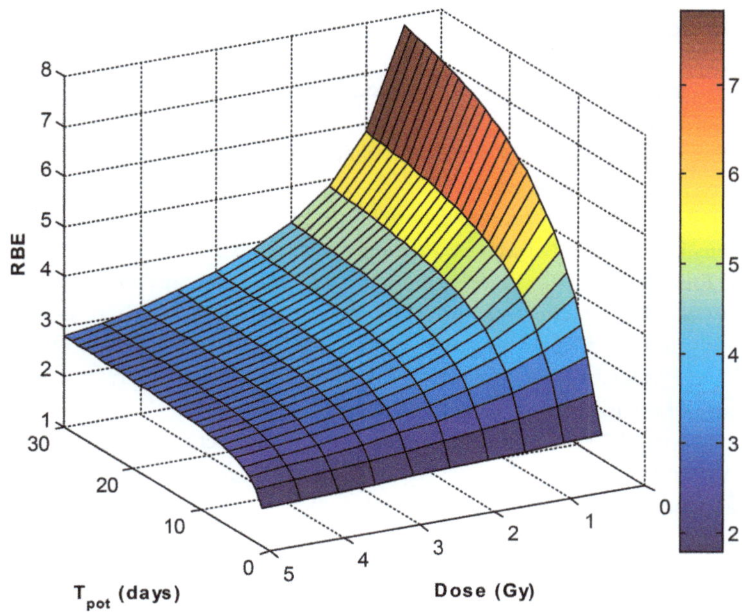

Figure 2.7. 3D plot of dose, T_{pot} and RBE for fast neutrons derived from data of Batterman *et al* (1981) the RBE relating to tumour recurrence times after treatment, using the inverse relationship with T_{pot} given above and an assumed cell loss factor of 90%. This may represent the worst-case situation for a proton beam, but would need scaling to lower RBEs for the more usual clinical proton beam, but may be useful for heavier ion beams.

References

Barendsen G W 1982 Dose fractionation, dose rate, and iso-effect relationships for normal tissue responses *Int. J. Radiat. Oncol. Biol. Phys.* **8** 1981–97

Battermann J J, Breur K, Hart G A M and van Peperzeel H A 1981 Observations on pulmonary metastases in patients after single doses and multiple fractions of fast neutrons and cbalt-60 gamma rays *Eur. J. Cancer* **17** 539–48

Bentzen S M and Baumann M 2002 The linear quadratic model in clinical practice *Basic Clinical Radiobiology* ed G G Steel (London: Arnold) p 135

Bopp C, Hirayama R, Inaniwa T, Kitagawa A, Matsufuji N and Noda K 2016 Adaptation of the microdosimetric kinetic model to hypoxia *Phys. Med. Biol.* **61** 7586–99

Britten R A *et al* 2013 Variations in the RBE for cell killing along the depth-dose profile of a modulated proton therapy beam *Radiat. Res.* **179** 21–8

Campbell I R and Warenius H M 1989 Radiation-induced cell death by chromatin loss. A model to explain the shape of low-linear-energy-transfer cell survival curves *Brit. J. Radiol.* **62** 338–43

Canney P A and Millar W T 1997 Biphasic cellular repair and implications for multiple field radiotherapy treatments *Brit. J. Radiol.* **70** 817–22

Carabe-Fernandez A, Dale R G and Jones B 2007 The incorporation of the concept of minimum RBE (RBE_{min}) into the linear-quadratic model and the potential for improved radiobiological analysis of high-LET treatments *Int. J. Radiat. Biol.* **83** 27–39

Chadwick K H and Leenhouts H P 1973 A molecular theory of cell survival *Phys. Med. Biol.* **18** 78–87

Curtis S B 1986 Lethal and potentially lethal lesions induced by radiation – a unified repair model *Radiat. Res.* **106** 252–70

Dale R G 1985 The application of the linear quadratic theory to fractionated and protracted radiotherapy *Brit. J. Radiol.* **58** 515–28

Dale R G 1989 Time-dependent tumour repopulation factors in the linear quadratic equations—implications for treatment strategies *Radiother. Oncol.* **15** 371–82

Dale R G and Jones B 1998 The clinical radiobiology of brachytherapy *Brit. J. Radiol.* **71** 465–83

Dale R G *et al* 2002 Practical methods for compensating for missed treatment days in radiotherapy, with particular reference to head and neck schedules *Clin. Oncol.* **14** 382–93

Douglas B G and Fowler J F 1976 The effect of multiple small doses of x-rays on skin reactions in the mouse and a basic interpretation *Radiat. Res.* **66** 401–26

Elsässer T and Scholz M 2007 Cluster effects within the local effect model *Radiat. Res.* **167** 319–29

Elsässer T, Kramer M and Scholz M 2008 Accuracy of the local effect model for the prediction of biological effects of carbon ions beams in vitro and in vivo *Int. J. Radiat. Oncol. Biol. Phys.* **71** 866–72

Fowler J F 1989 The linear quadratic formula and progress in fractionated radiotherapy *Brit J. Radiol.* **62** 679–94

Fowler J F 2008 Linear quadratics is alive and well: in response to Park *et al Int J. Radiat. Oncol. Biol. Phys.* **72** 957

Fowler J F 2010 21 years of biologically effective dose *Brit. J. Radiol.* **83** 554–68

Friedrich T, Scholz U, Elsässer T, Durante M and Scholz M 2013 Systematic analysis of RBE and related quantities using a database of cell survival experiments with ion beam irradiation *J. Radiat. Res.* **54** 494–514

Furusawa Y *et al* 2000 Inactivation of aerobic and hypoxic cells from three different cell lines by accelerated (3)He-, (12)C- and (20)Ne-ion beams *Radiat. Res* **154** 485–96 ; erratum 2012 *Radiat. Res.* **177** 129–31.

Hawkins R B 2003 A microdosimetric-kinetic model for the effect of non-Poisson distribution of lethal lesions on the variation of RBE with LET *Radiat. Res.* **160** 61

Hawkins R B 2009 The relationship between the sensitivity of cells to high-energy photons and the RBE of particle radiation used in radiotherapy *Radiat. Res.* **172** 761–76

Hopewell J W, Millar W T and Lindquist C 2012 Radiobiological principles: their application to γ knife therapy *Prog. Neurol. Surg.* **25** 39–54

Hopewell J W, Nyman J and Turesson I 2003 Time factor for acute tissue reactions following fractionated irradiation: a balance between repopulation and enhanced radiosensitivity *Int. J. Radiat. Biol.* **79** 513–24

Inaniwa T, Suzuki M, Furukawa T, Kase Y, Kanematsu N, Shirai T and Hawkins R B 2013 Effects of dose-delivery time structure on biological effectiveness for therapeutic carbon-ion beams evaluated with microdosimetric kinetic model *Radiat. Res.* **180** 44–59

Joiner M C, Marples B, Lambin P, Short S C and Turesson I 2001 Low dose hypersensitivity: current status and possible mechanisms *Int. J. Radiat. Oncol. Biol. Phys.* **49** 379–89

Jones B and Dale R G 2008 Radiobiological compensation for treatment errors in radiotherapy *British J. Radiol.* **81** 323–6

Jones B and Dale R G 2007 Repopulation effects *Radiobiological Modelling in Radiation Oncology* (London: British Institute of Radiology) pp 79–95

Jones B and Dale R G 2015 Radiobiology of high dose per fraction *Stereotactic Body Radiotherapy* ed A Gaya and R Mahadevan (Berlin: Springer)

Jones B and Morgan D A L 2007 Radiotherapy Fractionation *Radiobiological Modelling in Radiation Oncology* (London: British Institute of Radiology) pp. 70–1

Jones B and Sanghera P 2007 Estimation of radiobiological parameters and equivalent radiation dose of cytotoxic chemotherapy in malignant glioma *Int. J. Radiat. Oncol. Biol. Phys.* **68** 441–8

Jones B, Dale R G, Deehan C, Hopkins K I and Morgan D A L 2001 The role of biologically effective dose (BED) in clinical oncology *Clin. Oncol.* **13** 71–81

Jones B, Dale R G and Gaya A 2006a Linear quadratic modelling of increased late normal tissue effects in special clinical situations *Int. J. Radiat. Oncol. Biol. Phys.* **64** 948–53

Jones B, Carabe-Fernandez A and Dale R G 2006b Calculation of high-LET radiotherapy dose required for compensation of overall treatment time extensions *Brit. J. Radiol.* **79** 254–7

Jones B, Underwood T C, Carabe-Fernandez A and Dale R G 2011 Further analysis of fast neutron relative biological effects and implications for charged particle therapy *Brit. J. Radiol.* **84** 11–8

Kanai T, Furusawa Y, Fukutsu K, Itsukaichi H, Eguchi-Kasai K and Ohara H 1997 Irradiation of mixed beam and design of spread-out Bragg peak for heavy-ion radiotherapy *Radiat. Res.* **147** 78–85

Kase Y, Kanai T, Matsufuji N, Furusawa Y, Elsässer T and Scholz M 2008 Biophysical calculation of cell survival probabilities using amorphous track structure models for heavy-ion irradiation *Phys. Med. Biol.* **53** 37–59

Lea D E 1962 *Action of Radiation on Living Cells* 1st edn 1946 2nd edn 1962 (Cambridge: Cambridge University Press) ASIN: B0007JLWKQ

Marples B, Wouters B G, Collis S J, Chalmers A J and Joiner M C 2004 Low-dose hyper-radiosensitivity: a consequence of ineffective cell cycle arrest of radiation-damaged G2-phase cells *Radiat. Res.* **161** 247–55

Niemierko A 1997 Reporting and analyzing dose distributions: a concept of equivalent uniform dose *Med. Phys.* **24** 103–10

Okada T *et al* 2010 Carbon ion radiotherapy: clinical experiences at National Institute of Radiological Science (NIRS) *J. Radiat. Res. (Tokyo)* **51** 355–64

Park C, Papiez L, Zhang S, Story M and Timmerman R D 2008 Universal survival curve and single fraction equivalent dose: useful tools in understanding potency of ablative radiotherapy *Int. J. Radiat. Oncol. Biol. Phys.* **70** 847–52

Pop L A, Millar W T, van der Plas M and van der Kogel A J 2000 Radiation tolerance of rat spinal cord to pulsed dose rate (PDR-) brachytherapy: the impact of differences in temporal dose distribution *Radiother. Oncol.* **55** 301–15

Rossi H H and Zaider M 1992 Compound dual radiation action. I. General aspects *Radiat. Res* **132** 178–83 ; erratum 1993 *Radiat. Res.* **133** 274

Takahashi A *et al* 2014 Non-homologous end-joining repair plays a more important role than homologous recombination repair in defining radiosensitivity after exposure to high-LET radiation *Radiat. Res.* **182** 338–44

Thames H D, Withers H R, Peters L J and Fletcher G H 1982 Changes in early and late radiation responses with altered dose fractionation: implications for dose-survival relationships *Int. J. Radiat. Oncol. Biol. Phys.* **8** 219–26

Thames H D *et al* 2010 The role of overall treatment time in the outcome of radiotherapy of prostate cancer: an analysis of biochemical failure in 4839 men treated between 1987 and 1995 *Radiother. Oncol.* **96** 6–12

Wenzl T and Wilkens J J 2011 Modelling of the oxygen enhancement ratio for ion beam radiation therapy *Phys. Med. Biol.* **56** 3251–68

Wigg D R 2008 *Applied Radiobiology* (Madison, WI: Medical Physics Publication Corporation) ISBN10 1930524390

Wyatt R M, Beddoe A H and Dale R G 2003 The effects of delays in radiotherapy treatment on tumour control *Phys. Med. Biol.* **48** 139–55

IOP Publishing

Practical Radiobiology for Proton Therapy Planning

Bleddyn Jones

Chapter 3

Some important medical and surgical considerations, including clinical trials

3.1 Introduction

The radiotherapy physics textbooks do not normally consider the practical influence of important medical factors in decision making. A brief chapter is included here, which is designed to alert the non-medical reader (and medical readers embarking on their responsibilities) to some of the most important factors that could alert the clinician to change a treatment plan and influence the choice of the prescribed dose. Also, it is necessary to understand how radiation side-effect data is allocated, as well as how legal decisions as to whether radiation has caused tissue damage are taken. Medical physicists may be involved in correlating treatment plan dose distributions with patient outcomes and academic physicists may be unaware of the practical difficulties inherent in such analysis, as well as many of the terms in use.

3.2 Surgery

Radiotherapy is often preceded by surgery, which can vary in extent. Cancers can be partially removed (sometimes referred to as partial excision) or ‘debulked’, but surgery is often radical in intent (a term which implies that total excision or removal is attempted with the aim of achieving cure). Where there remains a risk of tumour regrowth due to residual cancer cells not removed by surgery, radiation can be given to reduce this risk, often with use of lower doses than would be given to an intact tumour. More uncommonly, surgery may follow the use of radiotherapy, for salvage of recurrent tumours or after tumours have been shown sufficient volume reduction to render them operable.

Surgery can be very prolonged, with retraction of normal tissues causing local tissue hypoxia. Dissection planes are cauterised using diathermy to stop bleeding, so there is inevitable devascularisation of normal tissues. Healing occurs by the process of scar formation, which itself can strangulate (or obliterate) fine blood vessels and

doi:10.1088/978-0-7503-1338-4ch3

nerves. Consequently, and since radiation cause late vascular damage, it is not surprising that the combination of surgery and radiation can lead to normal tissue injury of various kinds and to various degrees. In many body sites it is not advisable to give the full tissue tolerance dose of radiation after radical surgery because of the enhanced risk of complications.

Operating in previously irradiated tissues is also more hazardous, since the tissues are thickened due to the presence of scar tissue, with fragile blood vessels, years after radiotherapy, with prospects of delayed healing and wound breakdown accompanied by devitalisation of tissues.

In principle, there is a dose equivalent effect from surgery. For example, in Hodgkin's disease there appears to be a biological effective dose (BED) penalty of 17.7 $Gy_{[3]}$ in terms of abdominal radiation tolerance if the abdominal cavity has been explored (a laparotomy procedure), so the BED needs to be lowered by this amount in order to achieve the same effect (Jones *et al* 2006).

In BED terms, the effect of surgery can be expressed as a combined BED, consisting of the sum of the BEDs due to radiotherapy (RT) and surgery (SX) as

$$\text{BED(RT)} + \text{BED(SX)} = \text{BED(COMBINED)}.$$

Radiation treatment planning should always consider the nature and location of previous surgery, by close inspection of the operative notes and, where possible, direct conversation between the radiation oncologist and the surgeon regarding the approach taken and the true extent of dissection as well as the exact 3D position of a tumour before its removal, and what further tissue (such as a graft of fat or muscle) that may have been placed in or around that position.

It can be difficult to separate the true cause of complications when combined sequential treatments are given, but it is normal practice to classify a complication as being caused by radiation even if surgery has also been used, even if that complication could occur in the absence of radiation.

3.3 Cytotoxic chemotherapies

Similarly, chemotherapies of various kinds can be given before, during and after radiotherapy courses. Again, tissue radiation tolerance may be adversely affected and, in principle, a BED equivalent of chemotherapy can be estimated. For example the use of a relatively mild combination of three drugs called CMF (cyclophosphamide, methotrexate and 5-fluoro-uracil) is associated with a BED equivalent of 6.5 $Gy_{[3]}$ in terms of its contribution to chest wall fibrosis after radiotherapy for breast cancer. Similarly, the contributions are additive, so

$$\text{BED(RT)} + \text{BED(CHEMO)} = \text{BED(COMBINED)}.$$

3.4 Age and other medical conditions

It is increasingly recognised that late-developing vascular insufficiency causes many of the classical radiation side effects. Progressive obliteration of small vessels with changes in their morphology and function is characteristic, with subsequent deterioration of nutrition and oxygenation of tissues. The increasing prevalence of vascular disease with age, or the long-term effects of medical conditions that adversely influence tissue vasculature, such as hypertension, atherosclerosis, diabetes mellitus and other chronic inflammatory conditions, can also reduce radiation tolerance. Tissue damage, be it from direct radiation effects or indirect vascular effects resulting in chronic hypoxia, will cause scarring (or fibrosis). Medical conditions that are predisposed to scarring of tissues, such as rheumatoid arthritis and other autoimmune conditions, can also reduce tissue tolerance. Even older patients show a 3.6 $Gy_{[3]}$ change in tolerance for fibrotic effects after chest wall radiotherapy for breast cancer. The three examples (surgery, chemotherapy and age) can be found in Jones *et al* (2006).

The deposition of scar tissue is known as fibrosis; this can occur after surgery, radiation or other pathologies, and it is sometimes difficult to allocate the dominant cause when two or three of these factors co-exist. Severe or repeated infection can also cause tissue damage sufficient to cause fibrosis and, again, if there has been previous radiation exposure, then tissue breakdown can occur. Because of chronic hypoxia in 'heavily' irradiated tissues, there is a predilection for infection by anaerobic organisms (those bacteria that thrive in low oxygen tension conditions).

The very young may also exhibit reduced tolerances, with marked growth disturbance following radiation therapy. They also may live longer and so develop more late effects of radiation including malignant induction.

3.5 Interpretation of the case histories and literature

On the whole, if a patient has had radiation exposure in the past to an anatomical region where there is now a significant medical problem, the cause is usually allocated as radiation induced in publications. However, in medico-legal work, the cause has to be proven 'on the balance of probabilities', and, in instances where the radiation dose is considered to be too low and another competing pathology capable of such tissue changes is present, then negligence cannot be admitted. The situation with particle therapy is more complex, since further questions would need to be asked: 'could the assumed RBE or inaccurate Bragg peak-dose placement considerations be causative?'. This would need to be balanced by considering whether the treatment indications, its delivery and overall conduct would have been supported by a responsible group of practitioners (this is known as the Bolam principle), although this can sometimes be over-ridden by new scientific evidence. The author is not aware of any such cases at the present time, but they could arise in the future.

3.6 Clinical trials

These vary in intent from basic dose searching to identify efficacy and also side effects (or complications), when they are termed phase I and II. Phase III studies are normally randomised to compare a standard treatment with a new therapy and represent the highest level of evidence.

Prior evidence may be sufficient to suggest that only improvements can occur, in which case one-tailed probability levels can be used to test the null hypothesis, but normally two-tailed statistics should be used. In particle therapy, tumour control might be increased if the dose is escalated beyond that of the control, although in recent times the newer focused forms of photon therapy are quite capable of delivering the same high dose. The really vexing question is concerned with the two main uncertainties in particle therapy, namely dose or Bragg peak positioning accuracy and reliability, and the appropriateness of the relatively biological effect (RBE) value used. Dose positioning errors lead not only to dose changes but also linear energy transfer (LET) and consequently RBE changes, and can be expected to reduce tumour control and potentially increase normal tissue complications. The RBE uncertainty can produce either reduced tumour control if the prescription RBE exceeds that of the tumour being treated (since the dose will be reduced), or enhanced tumour control if the prescription RBE is less than that of the tumour (as the dose would be increased). Also, an incorrectly low allocation of RBE to a critical normal tissue could result in a higher risk of complications.

The terms local tumour control, disease free interval, or even biochemical control (for instances where tumour mass can be monitored by assay of a blood-borne chemical signal such as prostate specific antigen from prostate cancer) are sometimes used as endpoints, as well as patient survival, although the latter index can include local failure and salvage due to other treatments; time to first recurrence can sometimes be used to overcome the effect of later treatments. Actuarial statistics can be applied to all these measurements, where the index of survival is corrected for loss of patients due to death or lack of follow up due to a variety of reasons. Differences in outcomes can be tested by simple t-tests at specified time points or over a cumulative period of time by the log-rank test, but there are other alternatives.

For side effects, there is less uniformity since there are a number of different categorical scoring systems for each region/organ of the body. Crude percentages of a specific complication grading are often presented, but more sophisticated actuarial time series are preferred. Changes in the incidence of rare side effects can be quickly tested by simple Poisson statistics. For example, from a sample of 100 patients, if the expected incidence is 1% but a study finds an incidence of six, the probability is given by $\frac{1^6 e^{-1}}{6!} = 0.000\,52$. This is low and unlikely to be due to chance. If, on closer inspection, it is found that three of these patients with complications had another significant predisposing cause (such as prolonged and difficult surgery with devascularisation), and another had widespread vascular complications of diabetes, then only two clear-cut cases of toxicity due to radiation alone remain. The probability is then $\frac{1^2 e^{-1}}{2!} = 0.184$, which fails to be significant at the $p = 0.05$ level. In contrast, a 1% finding when the expected incidence is 6% yields a lower p value of 0.014 87, which passes the 5% significance test.

In most instances non-parametric statistical tests are normally used for the various categorical results. There are several risk related statistical endpoints in use, and in some instances quality of life may be fused with the survival endpoint (for example the percentage survival at five years in the absence of any documented grade 3 or 4 toxicity). Simulations of such an approach may lead to statistically significant findings with lower numbers of patients in a trial (Jones 2006). Textbooks of medical statistics should be consulted for further details.

3.7 Ethical issues

Much has been written about ethical considerations in particle therapy. There are enthusiasts who believe that treatment outcomes can only improve after particle therapy, but others insist that in many situations the randomised control study is essential to produce persuasive results that are independent of physician-related and other forms of bias (such as fitness to travel for therapy), which can influence outcomes, etc. The guiding precept in such situations is that there should be true equipoise between the standard and test treatments. Outstandingly good results from non-randomised studies may sometimes be convincing, but with any new treatment that contains intrinsic uncertainties, which may increase or reduce effectiveness, the randomised study should always be preferred. Further details on the opposing opinions are given in Sheehan *et al* (2014) and Jones *et al* (2014).

Randomisation would be one way of potentially deciding how the RBE should best be allocated. The present author has suggested randomly allocating certain classes of proton therapy patients to either the standard 1.1 RBE (for all tissues and doses) or alternatively to an RBE of 1 for radiosensitive tumours (as in children and lymphoma patients) with a higher RBE of 1.2 for any critical normal tissue. Such a trial would be important since at the present time it is difficult to separate whether failure of tumour control and/or increased toxicity is due to dose placement or to the RBE allocation. Radiosensitive tumours form a very special case since their RBEs are likely to be less than 1.1, so they may be under-dosed using the standard approach.

3.8 Mixed endpoints

It has been suggested that since particle therapy is mainly concerned with reducing tissue side effects, combined local control and toxicity endpoints could be used, as suggested already above. Such an approach is only valid if the tumour control remains constant or improves while side effects are reduced. In principle, such an approach—using for example an endpoint such as the percentage of uncomplicated recurrence-free patients, or say 'the percentage of recurrence-free with grade 1 toxicity', or even a disease-free survival corrected for adverse complications, which reflects useful quality of life (these are referred to quality adjusted)—can reach statistical significance for a smaller group of patients than would otherwise be the case, as in the example presented by Jones (2006).

3.9 The importance of follow-up

Cancer patients ideally require meticulous follow up appointments to assess their progress, with a detailed interview to check on symptoms, general examination and special examination of the relevant anatomical region. Various investigations including radiological techniques may be required. In general, the capture of relevant information is best when the patient attends the hospital where they were treated, and also a medical specialist in the area of the body that has been treated. Although effective questionnaires and regular communications with physicians and surgeons at remotely located hospitals can suffice, there can be under-reporting of the true incidence of side effects. It is often the case that when patients develop serious complications they seek help from other medical specialists who do not necessarily inform the original treatment centre. Particle therapy has been especially prone to this, since over the past thirty years many patients in the pioneering centres were referred from long distances, and often from different countries and continents. Modern IT communications can help in this respect, but there remain serious legal obstacles to the electronic transfer of medical data.

3.10 Publication bias

Under-reporting of medical research has been termed publication bias, and is now a well-recognised phenomenon. It remains to be determined as to why so many proton therapy centres have not published large and comprehensive reports of treatment effectiveness. This may be due to a number of reasons such as lack of resources for analysis, disappointing results, difficulties in analysis of complex patients and where follow up data may be incomplete. There may also be a tendency for journal editors and reviewers to reject what may be regarded as non-significant results or trials based on inadequate numbers. For example, the present author has advised acceptance of several submitted ionic beam papers for publication only to find that they had been rejected by other reviewers, sometimes when the research was unique and provided some 'proof of principle', although based on relatively few numbers of patients. It is vital that the results of new technologies applied in medicine are published with care. Another potentially misleading form of publication is where acute (early) side effects are only considered. These may be lower for newer radiation techniques and for particle therapy, but this does not necessarily mean that the late tissue complications (which may accrue over two to ten years or more) will also be reduced; indeed some categories of late complications may increase with time. Non-medical readers have to understand that acute reactions are those that occur mainly in surface or lining tissues (skin, gut, etc) where cell turnover times are rapid, whereas late reactions occur in more stable tissues due to the slow onset of late vascular damage. It is also pertinent to point out that late reacting tissues probably have a higher RBE than the acute reacting tissues, so some late reacting tissues may be more at risk following particle therapies if the RBE allocation has been incorrect. At low clinical dose per fraction, the RBE is inversely related to tissue α/β ratios; these ratios differ considerably between acute (10–30 Gy) and late reacting normal tissues (2–3 Gy).

3.11 Some future prospects

In order to overcome some of the above difficulties, it is suggested that data analysis and trial organisation should be arranged on a national or international, or even global basis. This is because each individual centre may treat only 500–1000 patients per year, but with say 20–30 different types of tumour indications. This means that individual centres may not accumulate patients quickly enough to produce sufficiently larger numbers of patients in each tumour class for analysis. National databases would accumulate data faster, and for rare conditions international data bases would be necessary. Countries with smaller populations could also contribute in this way. Furthermore, if national databases could identify individual cases with unexpected outcomes and then reconstruct the treatment plans using Monte Carlo-based simulations, for LET, inter-track distances and possible ranges of RBE, extremely useful information would be obtained for correlative purposes to identify which of the existing predictive models are the most suitable and in what clinical circumstances. In some respects, a single global centre could assist in checking such information from different national or international sources, using the latest Monte Carlo codes. In this respect, it may also be prudent to have quality assurance checks internationally or globally, as used to exist for fast neutron therapy trials. A single reference laboratory for particle therapy dosimetry would also be helpful, perhaps working in cooperation with existing international bodies like IAEA, the radiation division of the Bureau of Units and Measurements (Paris) and the particle physics simulation and IT expertise that exists in CERN (Geneva).

References

Jones B 2006 Implications of quality adjusted survival for clinical trials in radiation oncology *Brit. J. Radiol.* **79** 353–5

Jones B, Dale R G and Gaya A 2006 Linear quadratic modelling of increased late normal tissue effects in special clinical situations *Int. J. Radiat. Oncol. Biol. Phys.* **64** 948–53

Jones B, Howick J, Hopewell J and Liew S M 2014 Response to 'position statement on ethics, equipoise and research on charged particle therapy' *J. Med. Ethics* **40** 576–7

Sheehan M *et al* 2014 Position statement on ethics, equipoise and research on charged particle radiation therapy *J. Med. Ethics* **40** 572–5

Chapter 4

Treatment planning and further medical perspectives

4.1 Introduction

It is essential to describe the treatment planning process, while also including important medical and surgical factors that can influence radiation tissue tolerances referred to in the previous chapter. The non-medical reader will then be aware of reasons why a clinician may change a treatment plan and the prescribed dose. Also, it is necessary to understand how radiation side-effect data is allocated. Physicists may be involved in correlating treatment plan dose distributions with patient outcomes and academic physicists may be unaware of the practical difficulties inherent in such analysis.

The present author is aware that readers from different academic and vocational backgrounds require some basic explanations to help them better understand the context of some important practical dilemmas in radiation oncology.

4.2 Treatment planning processes

Radiation treatment planning must be appreciated as being only a part of the overall management of cancer, which consists of epidemiology, screening, diagnosis, clinical assessment including physical examination, pathological confirmation, imaging studies and determination of tumour grade and stage, and any other molecular based information that may influence treatment. Then the actual treatment may consist of surgery, radiotherapy, chemotherapy (cytotoxic, hormonal) and other approaches, either alone or in combinations determined by past experience and evidence. Especially important is the past medical history and age of the patient in the determination of an overall treatment strategy. These factors may also influence radiation tolerance, which is an assessment of how much radiation can the tissue tolerate, especially if other significant medical conditions co-exist. There may even

doi:10.1088/978-0-7503-1338-4ch4

be past exposures to radiation to consider, as well as the often deleterious effects of surgery. Some of these points will be expanded in later sections.

After the decision has been taken to proceed with radiotherapy, the overall treatment plan is commissioned by experienced radiation oncologists, who are medically qualified but trained to assess, care for and supervise cancer patients requiring radiotherapy. Their role includes defining the radiation treatment volume (s), and allocate an appropriate total dose and its fractionation (numbers of separate treatments) over a designated period of time (the overall treatment time). They may alert the patient and treatment staff about possible contingencies; important dose constraints for important normal tissue doses are defined, often as dose limits with or without volume considerations (expressed as the percentage volume of a tissue that may receive a recommended dose limit, for example the V-20, the volume which receives a minimum dose of 20 Gy).

Needless to say, these normal tissue constraints frequently influence the tumour dose, which is often termed the prescribed dose, analogous to a medical prescription of a drug, which has to be under the control of a qualified practitioner.

The subtleties that follow are important. The target volume definitions recommended by the ICRU are essential to understand. According to ICRU report 78 (2010), the visible tumour volume is called the gross tumour volume (GTV); the further volume that includes a margin for microscopic tumour extension is called the clinical target volume (CTV); and a further margin which compensates for daily shifts in positioning of the patient relative to the beam, movements during treatment and penumbral beam effects is called the planning target volume (PTV). In practice, the CTV is determined by clinical risk factors and can include a margin of 0.5–3 cm in all or some directions around the tumour, depending on its microscopic features (histological class, degree of differentiation and molecular markers). The CTV to PTV margin is often around a further one centimetre, but this can also vary with position in the body, the expectation of physiological movement, etc.

It is essential to know that the normal tissues situated around a cancer do frequently include infiltrating tumour cells, which cannot be detected on imaging studies. From clinical and pathological experience, it is known that certain types of tumour have more invasive propensities than others. Tumours considered to be relatively or completely benign have low-invasive or no-invasive potential, respectively. The CTV may also extend to include lymph node groups situated several centimetres away from a primary cancer, which may also harbour metastatic (or escaped) cancer cells. Thus the radiation oncologist faces the same dilemma as a surgeon as to how much tissue should be removed in the case of surgery, or 'sterilised' in the case of cancer cells with radiation, and what will be the effects on the patients if the normal tissue is rendered functionless. The surgeons are in a more fortunate position since they may use biopsies taken for immediate processing, which can advise them as to the effectiveness of their tissue dissections. No such aid is available for the oncologist, but surgical data backed by pathological evidence on distance of spread is sometimes available, although often incomplete, along with data from recurrence patterns after previous radiotherapy. This critical problem is to some extent made less uncertain by better definitions of the tumour edge, its disturbance of local vascularity such as permeability and blood flow changes that may reflect infiltration, as well as tumour metabolic activity by image fusion of CT (photon-based), MR (magnetic resonance-based) and PET (positron emission-based) scans, although the latter is prone to time-related fluctuations of perfusion of the delivered radioisotope, as can any perfusion-based study using contrast agents.

It is important to realise that the volume of normal tissue treated beyond a tumour can exceed that of the tumour itself. For instance, consider a spherical 2 cm diameter tumour with an additional 2 cm (linear) CTV 'margin' beyond it.

The volume of normal tissue is:

$$\frac{4}{3}\pi 3^3 - \frac{4}{3}\pi 1^3 = \frac{4}{3}\pi(8) = 33.\ 51\ \text{cm}^3,$$

whereas the tumour volume (the second term in the above equation) provides a volume of 4.19 cm^3.

As may be expected, this discrepancy increases further with tumour volume and the marginal distance used.

For a 4 cm diameter tumour, and for this particular situation, only a 1 cm margin is required, then the volume of normal tissue included is:

$$\frac{4}{3}\pi 3^3 - \frac{4}{3}\pi 2^3 = \frac{4}{3}\pi(19) = 78.59\ \text{cm}^3.$$

This exceeds the tumour volume (again the second term of the last equation), which is only 33.51 cm^3.

It is so often the case that for even small margins of 1 cm, the volume of normal tissue treated to the full dose will exceed that of the tumour itself, as shown in figure 4.1.

One can now appreciate that the normal tissue volume included in the full dose region can be substantial, as may a critical organ at risk (OAR) situated close to, but outside, the PTV margin. This may be especially relevant in particle therapy where a

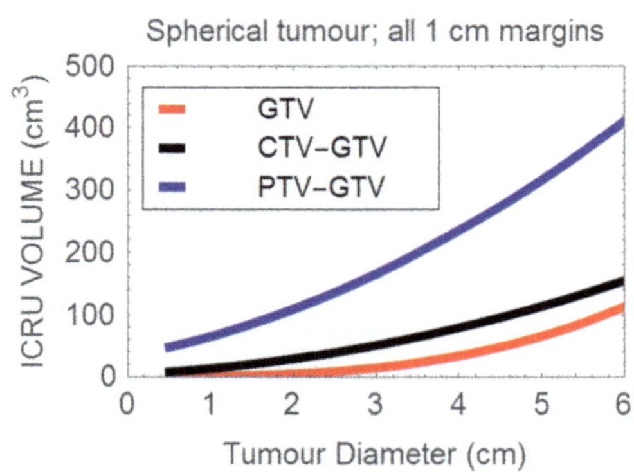

Figure 4.1. Plot of the three ICRU tumour volumes, with 1cm margins between the GTV and CTV, and between the CTV and PTV. Beyond the GTV, each volume is obtained by subtracting the neighbouring smaller volume, to show how much normal tissue is included between each volume.

prevalent relative biological effect (RBE) disadvantage due to raised linear energy transfer (LET) could in some instances be sufficient to overcome the degree of dose sparing of the OAR.

This clinical relevance of the normal tissues within the PTV and those close to it will depend on the functional importance of the tissue, and its ability to regenerate or not. The brain tissue is so often critical, but there are some areas that are relatively 'silent' in that they may involve areas that perceive sensory perception of limited importance, compared with precious areas that govern speech (the prefrontal cortex) or total vision (such as the optic chiasm). Other examples include the white matter tracts, which carry the neurological impulses that transmit motor signals, or in the reverse direction, direct sensory input to the central and processing brain. The criticality of the spinal cord, or optic chiasm, is a good example where severe radiation injury results in paralysis of movement of the limbs, and other diverse functions such as sensory loss, bladder and bowel incontinence, and impotence in males. In contrast, one can accept functional loss or depletion of some tissues such as skin (if treated to a critical depth and volume can effectively regenerate, and if sometimes severely damaged can be replaced by plastic surgery 'skin grafting' techniques); the liver, if healthy, is a tissue that can regenerate in three-dimensional space and so the margin becomes less relevant; if the tumour arises in muscle, then the surrounding tissue is muscle, which is radio-resistant compared to many other tissues. The primary tumours that arise in muscle and other connective tissues of the body are called sarcomas; depending on their location, most soft tissue sarcomas arising in muscle will not usually confer much motor deficit due to radiation beyond that already achieved by prior surgery. In this limited discussion it must be pointed out here that the sensitivity to an increment in LET with particle therapy may be greater in late reacting tissue, with slower cellular turnover kinetics, than in more acute reacting tissue that exhibits short cellular turnover times.

A further issue that can surprise some academic physicists is that, not infrequently, the primary target volume has no tumour on imaging, since it has been removed (or excised) surgically. There may be, at best, some imaging features of a tumour cavity and the operative path taken by the surgeon in some instances. The surgical 'tumour bed' and surrounding tissues may be harbouring residual tumour cells and require radiotherapy, usually to a lower dose than that required for an intact tumour, in order to reduce the chance of recurrence. In such cases the entire treatment volume is composed of normal tissues.

From the above considerations, it can be appreciated that the zonal definition of risk should be dependent on the tumour type and not the technique used to treat it. So often in radiotherapy, a new technique—perhaps with greater accuracy of photon delivery, etc—is accompanied by the temptation to use a reduced tumour margin. Disrespect by the physician for the true CTV may frequently lead to tumour-edge recurrences (that is tumours re-growing within their margin distance). Physicists may like to think of the tumour infiltration as being similar to electron probability clouds around the atomic nucleus.

Now that the CTV has been more fully considered, there is a further volume to be added, the PTV. This is necessary to account for daily set-up errors and movements of tissues during treatment such as due to breathing in the case of thoracic and upper abdominal organs where shifts of 1–3 cm are commonly encountered. This extra margin is not tumour dependent like the CTV, but is amenable to careful reductions by use of more meticulous and reproducible immobilisation techniques. Breath holding can be used, with physiotherapy training, but the more sophisticated technique of respiratory gating can be used, or the breathing tracked continuously with on-line feedback to change beam collimation or positioning. Particle therapy is not so forgiving as x-ray-based treatments as far as PTV dose compliance is concerned, because of the sharp dose fall-off associated with Bragg peaks, although these may be similar for some highly focused x-ray techniques, but there is always an exit dose with most rotational techniques. The highest risk of PTV non-compliance is with particle therapy pencil beams where there is always the possibility of the Bragg peak region being misplaced when the tumour and adjacent tissue is moving. Many research groups have addressed this problem and it was first tackled—using protons—by Japanese clinicians and scientists in the context of treating hepatomas in the liver around 25 years ago. Modern linear accelerators for x-ray treatments have cone beam CT scan facilities, which can assess the treatment accuracy on a daily or less frequent basis. Ultrasound techniques can also be used to check field limits and depths to a certain extent, together with the best possible radiographer skills. For conventional radiotherapy, the linac x-ray beam position quality assurance includes use of a portal imaging facility where the transmitted x-ray is captured beyond the patient and checked for accuracy against what was intended relative to the local bone anatomy. This is not possible during particle therapy because the energy deposition ends with the proton Bragg peak in most cases, or with the relatively small fragmentation tail of ions heavier than helium. Test doses of protons or ions at a much higher energy than required for the treatment could be used for this purpose (proton radiography) but the accelerator must then have higher energy-delivery capabilities. Protons can be used in this way in areas of the body with shorter tissue thickness such as the head and neck region, but not in the chest or pelvis without higher energy availability. However, the receptive detector technology is not yet commercially available. The considerable advantages of proton radiography include the fact that the absorbed dose throughout the thickness of the body is far less than that deposited by an x-ray beam, and also proton scatter by tissue is less than that from even megavoltage x-rays, so the tissue resolution is better.

4.3 The important interaction of RBE issues with the marginal target volumes

There is a real concern in CPT that the RBE is high in normal tissues as far as their late effects, mediated by vascular damage, are concerned, and especially so at low doses per fraction. This has the implication that if the RBE used in the tumour prescription is applied to all other tissues, and is lower than the RBE in the normal tissue of concern, then over-dosage could occur in the normal tissues included in the CTV and PTV. This may not be problematic in some clinical situations, such as in

Table 4.1. Dose status for each ICRU volume and likely clinical outcome changes produced by CPT compared with megavoltage photon radiotherapy, where ↑ is an increased dose, = is the 'equivalent' dose and ↓ is a reduced dose for the CPT. OTV is the volume of the body outside the PTV. Reproduced from Jones (2015a).

Dose status	Tumour control in GTV and {CTV + PTV}	{CTV + PTV} side effects	OTV side effects
GTV↑, {CTV+PTV}↑, OTV↓	Much better***	Worse*	Better
GTV↑, {CTV+PTV}=, OTV↓	Better***	Equal**	Better
GTV=, {CTV+PTV}=, OTV↓	Equal**	Equal**	Better
GTV=, {CTV+PTV}↓, OTV↓	Worse****	Better	Better or worse

*In some cases, reduced doses outside {CTV + PTV} may allow better radio-tolerance due to less late vascular insufficiency in the OTV. Also, this row condition may be acceptable in a non-essential tumour bearing tissue, or for a small volume of essential tissue with little risk of subsequent functional change.
**An equal outcome exists only if the RBE is correct.
***In some instances (the radiosensitive tumour classes), tumour control will be very high with x-rays and there will be no gain in tumour control from dose escalation using any form of radiotherapy. These tumours may also have a lower RBE than 1.1, in which case the tumour control probability will be reduced if an RBE of 1.1 is used.
****Reductions in dose to CTV and PTV can occur due to range uncertainties and tissue movement, sometimes exacerbated by the use of scanned beams. The clinician may decide to extend the PTV and/or increase the local dose, depending on the tissue contained in the PTV and as a 'trade-off' for gains in the OTV.

the liver where the surrounding tissue regenerates, or in examples such as connective tissue muscle or fat. The impact of a higher than anticipated RBE in a more precious tissue may be very important, as would be the case for anatomical locations such as critical areas of the brain, spinal cord, bowel, heart, etc. This situation can be expressed as an inequality for general simplification purposes, where if RBE[normal tissue, late] > RBE[prescription], then bio-effects may be greater.

This situation can be opposed by dose reduction, but this may carry the risk of under-dosage of infiltrating tumour cells. Another potential approach would be to increase the dose per fraction in order to reduce the late effect RBE, and its variation with the α/β ratio (discussed in chapter 2), although again care would be required to ensure an adequate tumour cell eradication dose.

It is instructive to look at a table of the ICRU volumes (table 4.1), as given also in Jones (2015b), and consider the implications of dose allocations to each volume with respect to later outcomes (tumour control and normal tissue complications). A significant change in the RBE may override the initial expectations. Readers may wish to construct their own table in particular treatment situations and vary the RBE assumption in each category.

4.4 Comparative planning studies

These studies compare CPT with x-ray techniques and, perhaps in the future, several different ion beams as well, in order to assess the dose distributions and to choose the best overall treatment plan. CPT patient selection will undoubtedly be based on such an approach, especially if coupled with comparative predictions of tumour cure probability (TCP) and normal tissue complication probability (NTCP). The Netherlands has an agreed system for doing this for proton therapy patient selection and its funding. Such an approach was considered to be an advance over the standard evidence-based medicine criteria, which demanded high level evidence from actual treatments, preferably in the context of randomised control trials. This type of exercise could produce erroneous predictions if the RBE assumptions are incorrect. Furthermore, NTCP models are notoriously unreliable and based on very simple physical representations of complex normal tissues so that volume related changes can be misleading. However, with reasonably correct allocations of the RBE, in principle it should be possible to predict photon equivalent doses and volumes, and relate these to existing photon data relating dose volume and risk. There is a real need to develop better systems in this respect, where well-verified LET-RBE models can be linked to dose per fraction effects and applied with existing low LET radiobiological parameters such as the α/β ratio.

The RBE issues in comparative planning studies can also apply to the tumour RBE. If, for example, the actual tumour RBE is less than the prescription RBE, then tumour under-dose could occur with possible reduced tumour control depending on the extent of under-dosage. In proton therapy RBE values lower than 1.1 are predicted in tumours known to be very radiosensitive to conventional (low LET) radiotherapy. These include the lymphomas and many childhood tumour types. Although their RBEs may be in the 1.03–1.05 range there would be a potential significant reduction in dose, which could exceed the ICRU recommendation of a 5% lower dose limit within the PTV. This may seem to be a trivial change, but it should be remembered that a 5% reduction of dose may prove to be significant if added to the already existing 5% guideline. This could also be compounded by small errors in dose delivery, Bragg peak positioning, etc, resulting in a recurrent tumour. In such circumstances it may be best not to use a conventional RBE for the tumour prescription, as long as the various normal tissue tolerances are respected. These tolerance doses are usually lower than the tumour prescribed dose for such radio-sensitive tumours, so there is a reasonable scope for such a policy in the treatment of these tumours where cure rates are high and quality of life is at a premium.

4.5 Trade-off situations in comparative treatment planning

A good example would be in the treatment of lung cancer, or oesophageal cancer, where an additional safety margin could be applied around a tumour to at least ensure 100% dose (rather than 95%) to cover the CTV and PTV. This would only be reasonable if there is an acceptable increase in the V-20 index (this refers to the volume of lung that receives over 20 Gy, the dose limit beyond which lung function can be impaired due to reduced gas exchange. The V-20 would need to remain less than what would be achieved with x-ray/photon techniques. The gas exchange loss following the alternative of surgical lobectomy or pneumonectomy should always be kept in mind, although in many examples of more unusual lung cancers the option of surgery has been rejected not only because of tumour stage (or size) but also because of the histological class. Similar arguments can be made for margins around primary liver cancers (hepatomas), pancreatic cancers, renal cancers (hypernephromas), or any situation where a small amount of additional parenchymal tissue may be sacrificed without major loss of function.

Another option in this type of situation is to express the V-20 or any other such index that might determine overall normal tissue tolerance in terms of another RBE-adjusted dose. It is important to ascertain whether the treatment planning system (TPS) is producing a 20 Gy physical dose volume statistic or an RBE-weighted one. If the latter is assumed and, for example, if the late effect RBE of lung tissue in the mid-spread-out Bragg peak was considered to be 1.3 rather than the 1.1 used in the TPS, then one could consider the V-20 to be actually a V-23.64 (where 23.64 is $20 \times 1.3/1.1$).

4.5.1 Changes in the treatment plan

During radiotherapy the treatment plan and dose prescription is occasionally modified, and for good reasons. These include:

1. Severe acute reactions causing medical complications and other causes of treatment interruption such as accelerator breakdowns, inter-current illness, failure to attend due to inclement weather or other causes.
2. Errors in treatment delivery (sometimes manifesting as severe acute reactions if over-dosage occurs and a lack of acute reaction at an appropriate time with under-dosage). These may occur due to dose-calculation errors, or be due to changes in body anatomy such as weight loss or weight gain, which can result in under- or over-dosage and which may require correction. Rarely, the treatment field position is incorrect and so a compensatory plan is necessary. The remaining treatment may require a recalculation of the dose per fraction using biologically effective dose (BED) equations.
3. Further information such as examination of the patient or new imaging studies reveal tumour spread or its expansion from whatever cause (e.g. cystic degeneration or bleeding, with fluid accumulation within the tumour).
4. Adaptive therapy defines a more flexible approach for modifying treatment plans on a deliberate basis according to information gained by serial imaging such as tumour shrinkage rates, shifts of normal tissue, etc.

More detailed examples of the above are given in the chapters already referred to, with some worked examples that illustrate the additional pitfalls that exist with charged hadrons in BED equations. It should also be noted that particle beams are more sensitive to changes in body anatomy, due to misplacement of Bragg peaks, which may change the LET, and the RBE, as well as dose. X-ray/photon-based treatment is much more forgiving, not only in physical dose placement, but also in radiobiological terms, as will be shown later.

4.6 How to accommodate assumed errors in the RBE

The relationships between LET and RBE will be discussed in chapters 8 and 9, but in the present context it is assumed that there is a lack of information, or a range estimate, as to what the most appropriate RBE might be. If models that link LET and RBE are not used, but if changes of RBE are to be respected in critical tissues such as the central nervous system (CNS), the following approach might be used.

Essentially, the easiest way of taking variations in the proton RBE into account will be to assume worst-case scenarios in the critical normal tissue of interest, and to use a larger RBE than that used in the planning process, where at the present time all RBEs are 1.1. If the normal tissue tolerance constraint is, say, a dose of 54 Gy, and the plan provides 51.5 Gy-RBE, then if the RBE was as high as 1.2, then the adjusted dose would be 1.2/1.1 × 51.5 = 56.2 Gy-RBE, which exceeds tolerance; if the RBE was even higher at say 1.25 or 1.3 then the respective Gy-RBE values would be 58.52 and 60.87, which significantly exceed the constraint and carry an enhanced risk of toxicity.

Such an approach may be all that is necessary but it does not include the tumour RBE. A more sophisticated approach might be to use both tumour and normal tissue RBEs, as in the approach taken by Jones *et al* (2011). As throughout this book, suffixes of L and H are used for low and high LET radiations, respectively. If the tumour dose per fraction at the prescription point is z_H for high LET and z_L for low LET radiations, then since RBE is the ratio of the low to high LET single doses (or dose per fraction) necessary for a given biological iso-effect, then the required value of z_H in the high LET case may be found from the RBE specific to the tumour volume (RBE_{tum}) by:

$$z_H = \frac{z_L}{RBE_{tum}}. \tag{4.1}$$

Equation (4.1) represents the iso-effective dose conversion required when the fraction number is the same for both the high and low LET case. In most forms of radiotherapy, but especially in CPT, a degree of physical normal tissue sparing is expected for normal tissues outside the PTV, such that the physical dose (d_H) to a critical region of normal tissue (e.g. the maximum or modal dose given to a clinically relevant normal tissue point or volume according to local anatomical constraints) is given by:

$$d_H = z_H \times P_H \tag{4.2}$$

where P_H is the physical sparing factor of the high LET radiation, which is the fraction of the tumour dose (z_H) delivered to the normal tissue. Ideally P_H should be as low as possible. For low LET the associated sparing will likely be less, i.e. the physical sparing factor (P_L) will be greater than P_H. For example, if a vertebral tumour is given a dose z_L by low LET or z_H by high LET, the dose at the surface of the spinal cord a few millimetres away might be $P_H \times z_H = d_H$ in the high LET case and $P_L \times z_L = d_L$ in the low LET case.

Using equations (4.1) and (4.2), the normal tissue high LET dose per fraction, d_H, can be converted back to an equivalent low LET dose per fraction, (d_{Leq}), by use of the normal tissue RBE (RBE_{norm}), to give:

$$d_{Leq} = z_H \times P_H \times RBE_{norm}. \tag{4.3}$$

Then by substituting z_H from equation (4.1):

$$d_{Leq} = z_L \times P_H \times \frac{RBE_{norm}}{RBE_{tum}}. \tag{4.4}$$

For the low LET radiation the normal tissue dose is:

$$d_L = z_L \times P_L \tag{4.5}$$

where P_L is the physical sparing factor for the low LET radiation and $P_H < P_L$ in most circumstances. Combining equations (4.4) and (4.5) and rearrangement of terms leads to:

$$\frac{d_{Leq}}{d_L} = \frac{P_H}{P_L} \times \frac{RBE_{norm}}{RBE_{tum}} \tag{4.6}$$

and re-written as:

$$S_0 = \frac{P_H}{P_L} \times \frac{RBE_{norm}}{RBE_{tum}} \tag{4.7}$$

where $S_0 = d_{Leq}/d_L$, represents the combined physical and radiobiological sparing. S_0, which could also be called the biological sparing advantage factor, needs to be as small as possible for particle therapy to have advantage over conventional x-ray therapy. It is clear that the precise value of S_0 will not only depend on the physical dose sparing ratio, but will also be highly sensitive to errors made in estimating the RBE for both tumours and normal tissue, especially as the RBEs are themselves dependent on fraction size, intrinsic low LET radiosensitivities and tissue cell kinetics. To give an S factor 'inclusive of error terms' (S_I), the errors in the RBE values may be incorporated as multiplicative terms on the normal tissue and the tumour RBEs values (defined as $Error_d$ and $Error_z$, respectively), resulting in:

$$S_I = \frac{P_H}{P_L} \times \frac{RBE_{norm}}{RBE_{tum}} \times \frac{(1 \pm Error_d)}{(1 \pm Error_z)}. \tag{4.8}$$

The error term can be simply treated as a factor from 0 to 1 or as a percentage change. For example, if the recommended RBE weighting factor is 3, while actually it is 1.5 in a particular tissue of interest, this represents a 50% error, whereas if the actual RBE is 6, there is a 100% error. In equation (4.8), such an error of 50% would be included as a factor 0.5, an error of 30% as 0.3 etc.

In this way it can be seen that the errors in the RBE, if in the wrong 'direction', can impose a higher demand on treatment planning. For example, applying 20% errors to equation (4.8):

$$S_{w,20\%} = \frac{P_H}{P_L} \times \left(\frac{RBE_{norm}}{RBE_{tum}} \times \frac{1+0.2}{1-0.2}\right)$$

or

$$S_{w,20\%} = \frac{P_H}{P_L} \times \left(\frac{RBE_{norm}}{RBE_{tum}} \times \frac{3}{2}\right).$$

By substituting S_0, the S value assumed prior to the consideration of the RBE errors (equation (4.7)) into equation (4.11):

$$S_{w,20\%} = S_0 \times \frac{3}{2}. \tag{4.10}$$

Thus for S_0 to be maintained in the case of $S_{w,20\%}$, the high LET physical dose sparing factor, P_H, would have to be reduced by one third:

$$S_{w,20\%} = \left(P_H \times \frac{2}{3}\right) \times \frac{1}{P_L} \times \left(\frac{\text{RBE}_{\text{norm}}}{\text{RBE}_{\text{tum}}} \times \frac{3}{2}\right) \Rightarrow S_{w,20\%} = S_0. \tag{4.11}$$

Further discussions and examples are given in Jones *et al* (2011).

4.7 The product of LET and dose

It has recently been suggested in international meetings that that the product of LET and dose can be used to replace the RBE. This product cannot fully represent the likelihood of biological effects, especially since the RBE varies inversely with dose itself. The product's physical units are not meaningful (energy released/distance × energy absorbed/mass) and is seen to be insufficiently discriminating in figure 4.2, where the product is plotted against the equivalent dose in 2 Gy fractions for an iso-effect based on the BED concept applied to the proton cell survival data of Britten *et al* (2013), using the proton LET model described in chapters 8 and 9. The large phase space occupied shows that the use of the product, which may produce erroneous predictions, despite the overall trend showing a positive correlation. Further details are given in Jones (2017).

The same can be deduced from the plots of the product against surviving fraction for ionic beams in figures 4.3(a–c), where the overlap is often significant, even in the case of substantial changes in the surviving fractions.

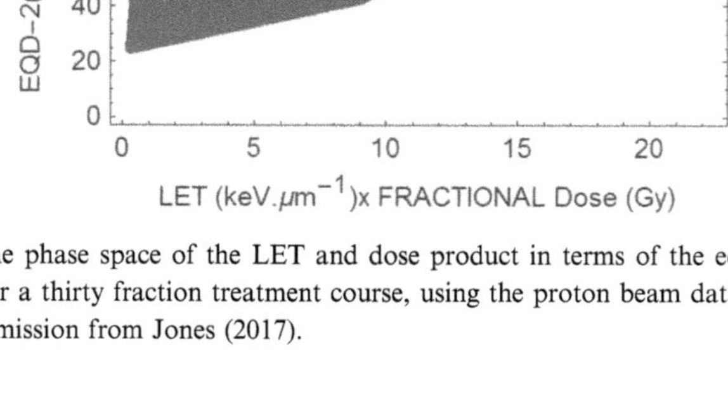

Figure 4.2. Plot of the phase space of the LET and dose product in terms of the equivalent dose in 2 Gy fractions (EQD-2), for a thirty fraction treatment course, using the proton beam data by Britten *et al* 2013. Reproduced with permission from Jones (2017).

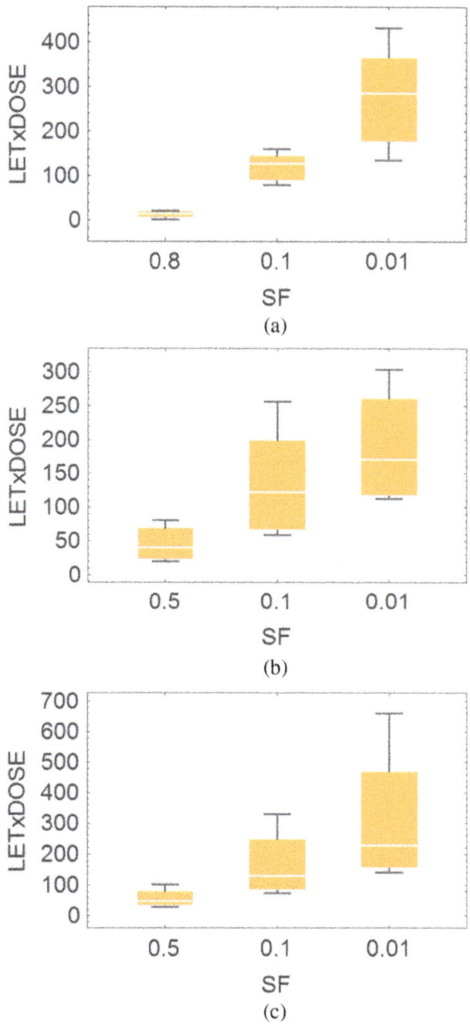

Figure 4.3. Box and whisker plot of Dose × LET product for different surviving fractions (a) Data of Barendsen (1968): human T cells exposed to deuterons and helium ions. (b) Weyrather *et al* (1999) data for C ions using CHO cells. (c) Weyrather *et al* (1999) data using V-79 cells. Reproduced from Jones (2017).

References

Barendsen G W 1968 Responses of cultured cells, tumours and normal tissues to radiations of different linear energy transfer *Curr. Topics Radiat. Res.* Q **4** 293–356

Britten R A *et al* 2013 Variations in the RBE for cell killing along the depth-dose profile of a modulated proton therapy beam *Radiat. Res.* **179** 21–8

ICRU Report 78 2010 *Prescribing, Recording, and Reporting Proton-Beam Therapy* (Bethesda, MD: NIH)

Jones B 2015a A simpler energy transfer efficiency model to predict relative biological effect (RBE) for protons and heavier ions *Front. Oncol.* **11** 184

Jones B 2015b Towards achieving the full clinical potential of proton therapy by inclusion of LET and RBE models *Cancers* **17** 460–80

Jones B 2017 Clinical radiobiology of proton therapy: modeling of RBE *Acta Oncolog*. **18** 1–5

Jones B, Underwood T C and Dale R G 2011 The potential impact of RBE uncertainty on charged particle treatment prescriptions *Brit. J. Radiol.* **84** S61–9

Weyrather W K, Ritter S, Scholz M and Kraft G 1999 RBE for carbon track-segment irradiation in cell lines of differing repair capacity *Int. J. Radiat. Biol.* **75** 1357–64

Bleddyn Jones

Chapter 5

Historical development of radiotherapy including what was learned from fast neutrons

5.1 Introduction

The following brief description of the historical development of radiotherapy is necessary in order for readers from some backgrounds to understand the present dilemmas facing decision makers. It acts as an interlude, but also explains some of the problems associated with technical development (most of which has been beneficial), and the example of fast neutrons contains important lessons that are relevant to particle therapy today.

5.2 A brief synopsis

The discovery of x-rays and natural radioactivity were soon followed by their application in medicine. Diagnostic imaging became widespread within a few decades; radiotherapy of cancer, as an alternative to surgery, became increasingly important, and remains so. Technical developments were mainly based on the achievement of higher photon energies, which increased the range or tissue depth while also, for megavoltage photons, reducing the skin or entry dose before attaining full secondary electronic equilibrium. Improvements in dosimetry and computational interactions yielded further advances in medical diagnosis, including nuclear medicine (radio-isotope emissions), computerized tomography (CT scans), positron emission applications, as well as imaging advances from other branches of physics (MR imaging and ultrasound). All these imaging techniques have been used to further improve radiotherapy targeting and it became possible to superimpose radiation depth dose curves in 3D on the relevant scan or even combination of fused scan images.

doi:10.1088/978-0-7503-1338-4ch5

During the second World War an important advance was made by Frank Ellis who realized during an air raid[1] that the local beam intensity could be varied. This was achieved by introducing a wedge shaped medium (initially wood and later metal) to attenuate the beam along its axis in order to produce more homogenous doses around a target when multiple beams were used, and where the anatomical shape of the body would distort the regular profile of the beam. They became known as wedge filters and compensators were a further extension where small areas of the beam were made to pass through tailor made thicknesses of metal that were based on the thickness of tissue to be traversed.

Naturally occurring radioisotopes were used in an increasingly controlled manner with excellent results if the radioactivity could be concentrated within a cancer; the use of rare and manufactured isotopes and various methods of enhancing their localisation continue to be developed. The best example is thyroid cancer, which, although rare, has distinct subtypes each with different propensities to trap iodine. Cure rates are very high for the avid iodine trapping cancers, even when these have spread to distant parts of the body. Sealed radioisotope sources also became important for applications where the sources could be placed within or close to many forms of primary cancers and remain in use.

During the 1950s and 1960s external radiation beams progressed to the megavoltage energy range by use of 60-cobalt units and later increasing use of linear accelerators, after the potential of the cavity magnetron had been fully realized for radar applications, to an extent that the already existing cyclotron accelerators were only rarely used for proton acceleration onto beryllium targets to produce fast neutrons for clinical studies.

Along with the computing advances mentioned above, during the 1990s it became possible to shape each individual beam in order to match the clinically defined target by introducing variable strips of metal in the accelerator collimation system. This became known as conformal radiotherapy. A UK randomized clinical trial showed that normal tissue side effects were reduced due to the large reduction in normal tissue volume being irradiated to a high dose but without a reduction in prostate cancer tumour control with long term analysis (Dearnaley *et al* 2014). The beam control possibilities were improved further by using the multi-leaf collimator to vary the beam intensity along its profile: this became known as intensity

[1] Ellis was asked by his physicist to go to the deeper end of the air-raid shelter, which was better protected, since there were a greater number of sandbags above that section than over the entrance region; it was explained that the roof was effectively wedge-shaped, the thin end being over the entrance. Ellis remarked that they should use similar principles in radiotherapy to overcome the problem of curved anatomical surfaces, and they started drawing diagrams of what was necessary. So, in some respects, Goering's Luftwaffe bears some responsibility for this important development. Ellis told this to the present author at a special dinner given by the British Institute of Radiology to commemorate his 100th birthday, and has never been published.

modulated radiotherapy, which further improved the degree of conformity[2] to the defined target volume. Although the UK was slow to adopt this technique, randomized studies were also performed to confirm the improved normal tissue sparing of defined organs such as the parotid gland while treating head and neck cancer (Nutting *et al* 2011). In order to achieve the best available CI around the immediate vicinity of the target volume, it was necessary to use more individual beams, and with the differential attenuation used the net effect was to increase the amount of tissue exposed to medium or lower doses, with some potential for causing more subtle later effects such as cancer induction or late vascular effects over time periods of 5–30 years. More recently, developments include robotically controlled small linacs, offering rapid changes in beam direction and intensity modulation of small beamlets, which can further improve the CI and can be used with more precise body immobilization techniques and state-of-the-art image guidance techniques. These treatments, depending on the dose distributions achieved, can be given in fewer treatments (hypo-fractionation) of even in a single session (often referred to as radiosurgery).

More recently, there has been an expansion in cyclotron or synchrotron acceleration to deliver protons and light ions for cancer therapy. These positively charged particles have Bragg peaks whose tissue depths can be controlled by good energy selection coupled with detailed imaging, so that energy deposition can occur in the selected cancer volume and its immediate surroundings.

In comparison with the best available photon beams, charged particles offer reduced entrance doses and no direct dose beyond the target. However, when more than one beam is used and using pencil (or raster) intensity modulated beams, good to excellent tumour conformity can be achieved. However, the degrees of freedom of angular selection are considerably reduced, and even if expensive gantries are available the degree of conformity in the immediate tumour region may not be as good as with some photon techniques, but the integral dose (that is the dose over the entire part of the body being irradiated) can be substantially reduced. Thus, the medium to low dose volume is reduced and more tissue tends to receive no direct radiation, which would be the case with all modern photon techniques. One drawback is that skin sparing can be suboptimal compared to linacs or electrons when protons are used for volumes near to the skin surface.

Both radioisotopes and charged particle beams rely on preferential radiation dose distribution, respectively due to sharp inverse-square dose fall-off with distance and the Bragg peak effect.

[2] The conformity index (CI) refers to a single-figure assessment of the merit of a planned dose distribution compared to an ideal situation where the radiation dose is completely uniform (or where the target volume dose is totally achieved), with or without a desired reduction of dose to further volumes of normal tissue. There are many variants of CI in use for this purpose, some of which may be misleading in some situations, since they tend to be defined with certain techniques in mind. The reader should consult widely before selecting any one particular system. For particle therapy the situation is more complex due to linear energy transfer (LET) and dose inhomogeneity, and the main advantage is dose reduction or no direct dose in some tissues, reflected more by the integral dose in the region outside the target volume limits.

Research studies have shown that the biological effects of radiation vary not only with dose but the dose rate, the degree to which the dose can be split in time (fractionation), the chemical environment of the cells (some chemicals protecting and others sensitising radiation by influencing the yield of free radicals), and also the 'quality' of the radiation. The latter refers to the LET characteristics of a radiation, which depends not only on the nature of the radiation (e.g. photon or hadron), but also on its energy (lower energies confer higher LET) and nuclear charge.

A major question that remains to be completely settled is whether the increase in LET may be disadvantageous to some normal tissues situated very close to the tumour, depending on the anatomical region being treated. This is taken further in later chapters (7–12).

5.3 Neutron therapy

Many current authorities tend to be dismissive of the history of fast neutron therapy because of its disappointing clinical outcomes, often interpreted as being 'bad press' for radiotherapy in general and especially for any form of particle therapy emerging from a cyclotron. This is unfair for several reasons, as shown in the outline provided below with special reference to their physical properties (uncharged and similar in some respects to photons) but with high LET features. Fast neutrons cause most of their ionization by forming recoil protons as well as some nuclear fragments, so their radiobiological features, including their relative biological effect (RBE) values will be similar to those of a proton beam in the Bragg peak region where the LET and dose increase substantially. This fact has not been sufficiently appreciated within the proton therapy community until recently. Indeed RBE values in the fast neutron range have been found at the end of spread out Bragg peaks in the human lung (Underwood and Paganetti 2016, Underwood 2017).

Fast neutron therapy of cancer started optimistically on the basis of limited *in vitro* experimental evidence of more efficient cell killing per unit dose, but unfortunately the clinical results were disappointing. The history of neutron therapy illustrates the frustrations encountered when partial scientific knowledge is used in an attempt to improve the treatment of a complex biological condition such as cancer; the sterilisation of cancer cells in a laboratory experiment is considerably easier than the elimination of a malignant tumour situated close to essential organs/tissues of the body. In the context of future radiotherapy developments, particularly the use of proton and ion beam therapy, this history is highly relevant.

Scientists such as Gray *et al* (1940) had shown that it was possible to achieve the same level of biological effect with a lower dose of neutrons than with γ- or x-rays. Such a difference is quantified by the RBE, the ratio of the doses of the two forms of radiation required to produce the same biological effect. Fast neutron RBE values of 1.5–5 were found in a variety of biological systems (bacteria, plants and transplanted animal cancers). The immediate inference was that neutrons would be ideal for cancer therapy, yet the first human experimental treatment in the USA showed marked toxicity, because the relationship between the exposure dose and the RBE had not yet been identified. Gray remained sceptical, and realized that in the three spatial dimensions encountered within the human body, neutrons could cause more problems because of the RBE; he knew that the RBE varied inversely with dose, so that dose fall-off with distance along a neutron beam would inevitably be accompanied by higher RBE values in normal tissues beyond any cancer target. His opinion had devastating personal[3] and other organizational consequences for radiobiology and radiotherapy, especially in the UK (Jones and Hendry 2017). Gray believed that neutrons were an important tool for research for the investigation of high LET effects, but not necessarily in treatment. He was eventually proved correct and the wealth of fast neutron experimental data (much of it performed in the UK) probably provides the best insights into high LET phenomena, especially the inverse dose per fraction effect on the RBE, especially for late reacting tissues rather than for acute tissue effects.

Further interest arose because of the discovery that high LET radiations, e.g. fast neutrons, with increased clustering of ionization events along micrometre distances of their tracks, are less dependent than x-rays on the presence of oxygen to produce cell death (oxygen essentially amplifies low LET ionisations by increasing the yield of reactive free radicals in solution). The previous work of Gray and others had shown that many cancers contained zones of very low oxygen tension, which were considered an important cause of radio-resistance. To overcome this problem, high pressure oxygen (HPO) was used in experimental radiotherapy, with impressive results. In the UK, it had been developed in Hammersmith; the initial results in animal experimental systems were impressive (Gray *et al* 1953). In clinical practice HPO had many disadvantages, since patients had to be placed within HPO tanks or chambers, and there was no overall improvement in patient survival although some tumour types were better controlled (Henk *et al* 1977). An attractive alternative to HPO was the use of cyclotrons to accelerate protons—to around 20 MeV or higher—colliding with beryllium targets to produce fast neutrons with high LET properties and so a reduced cell killing oxygen dependency within cancers. It was argued by the neutron enthusiasts that the HPO chambers would no longer be necessary.

The UK MRC funded three important sequential projects to investigate fast neutron therapy.

1. At Hammersmith, clinical studies were conducted with initial promise using a geometrically limited fixed horizontal beam. However, despite clear evidence that the neutron RBE was inversely related to dose per fraction

[3] Dr Constance Wood, Medical Director of Radiotherapy, dismissed Gray from his post as Director of Radiotherapy Physics at the Hammersmith Hospital, where important ground breaking work had been done on describing the cell cycle and other bio-medical advances. Wood had a distinguished clinical background and had used some of her personal inheritance to jointly fund the development of the first linear accelerator in Europe, in partnership with Metropolitan Vickers who manufactured radar and other defence-related equipment. She later regretted her decision, based on her own conviction that neutrons were the way forward for radiotherapy. In order to salvage Gray's career, a new laboratory was created for him on the Mount Vernon Hospital site in Northwood, London and it even bore his name. The rest is history since the Gray Laboratory, funded by some UK cancer charities, the MRC, and with impressive contributions by The Scott of Yews Trust, has become the World-leading laboratory for radiobiology research (Hendry and Jones 2017). It is interesting to note that of the long list of distinguished scientists who had been Fellows of Trinity College Cambridge, only Newton and Gray have SI units named after them.

in a wide variety of animal tissues, the clinical dose prescriptions used a fixed RBE. So, the dose plan took no account of the increase in RBE in normal tissues beyond the tumour, which received lower doses than those prescribed to the tumour. Initial attempts at randomised trials involved control patients treated with x-rays or cobalt beams at other hospitals without defined protocols so that a wide range of doses, including some that were unsuitable, were used. Suffice it to say that much was learned on how to conduct cancer trials properly.

2. At Edinburgh, stricter 'in-house' randomized trials were carried out comparing megavoltage x-rays (with superior tissue penetration) and relatively poorly penetrating fast neutrons, but for both radiation classes the beams could be rotated on a gantry. However, the tumour control rates were disappointing and were accompanied by enhanced normal tissue toxicity.
3. At Clatterbridge (Liverpool), an extended fast neutron energy (obtained using 64 MeV protons) produced depth doses equivalent to 5 MeV x-rays in randomized trials, some of which were jointly undertaken with Seattle (USA).

Taken together, these trials showed that neutrons conferred no clinical advantage, as summarized by Duncan (1994). In other countries, relatively low energy neutrons had been tried without recourse to formal trials and with little convincing success, although one small randomized trial reported by Laramore *et al* (1993) showed benefits for neutrons in the control of unresectable cancers of the parotid gland, but it is possible that a higher dose of x-rays or electrons in the control arm might have produced the same result by matching the actual tumour RBE, which was probably higher than that used in the prescription since most parotid tumours are slow growing. Relatively superficial air sinus cancers were also thought to be better controlled, although there was always concern that neutrons were particularly damaging to the tissues of the underlying brain, where it had been identified that the RBE was around 5 rather than 3.

In retrospect, neutron therapy failed to match its original promise for the following physical and radiobiological reasons.

- Routine absorbed dose computations did not include the highly efficient neutron capture by hydrogen resulting in higher energy release in hydrogen-rich tissues such as brain white matter and fat, which surrounds most important organs and is closely associated with their blood supply.
- Dismissal of the well-established finding of RBE variations in different tissues and its important increase with a falling dose, which mitigates the effect of a reduction in physical dose beyond a cancer.
- The appreciation that the RBE also varies with cell proliferation rate, so that slow-growing cells have higher values. It is the slow-growing cells that make up the majority of normal tissues and contribute to the severe tissue damage at extended time periods after irradiation.

The considerable literature on neutron radiobiology was summarized by Bewley (1989) and remains relevant to proton therapy. More recent mathematical modelling, which included the RBE within biological effective dose (BED) equations, showed that neutrons would only have improved therapeutic results in the case of very superficial cancers with little normal tissue coverage (Dale and Jones 1998).

5.4 More recent developments based on neutron studies

Fast neutron radiobiology studies had already shown high RBE values that varied inversely with dose, reductions in OER and cell cycle phase dependency, with greater fraction insensitivity (Field 1976). Extended analysis of the fast neutron experiments at Hammersmith and Clatterbridge have continued in more recent times, providing many informative reports, such as in Carabe-Fernandez *et al* (2007, 2010), Jones (2010) and Jones *et al* (2011), which model the well-recognized inverse dose per fraction and the RBE effects (using the RBE_{max} and RBE_{min} concepts, defined in chapter 2) in different tissue classes represented by their characteristic α/β ratios. For example, early reacting tissues such as the acute oesophageal mucosal reaction show almost no change in the RBE with dose per fraction, whereas later reactions in various tissues such as skin, lung and kidney show greater changes in the RBE with dose per fraction. In accord with the linear-quadratic model theory (see the appendix at end of this chapter), RBE_{max} is inversely related to the α/β ratio (see figure 5.1(a)) while RBE_{min} is directly proportional to the square root of α/β (see figure 5.1(b)). By using these relationships, for the most critical low α/β (late-reacting) tissues the RBE at low dose is highest, but becomes the lowest RBE at high dose when compared with more rapid proliferating bio-systems with high α/β ratios, which have a 'flatter' response, as shown in figure 5.2(a).

By re-scaling figure 5.1 to match a 1.1 proton RBE, for $\alpha/\beta = 10$ Gy (the value of the predominant jejunal crypt *in vivo* assay used to achieve this RBE in mid SOBPs), it is possible to predict as a first approximation what the RBE values would be at other doses and for different tissue types, as shown in figure 5.2(b). Again there is a cross-over effect where for large dose per fraction, the RBE values are lowest in the tissues with low α/β values, the reverse of that found at low dose per fraction. At 2 Gy per fraction, the α/β of 2 Gy characteristic of central nervous tissue is around 1.2. It is noted here that such a cross-over effect is reliant on using the RBE_{max} and RBE_{min} limits; other radiobiological models of proton and ion beam therapy that do not use these limits will not predict such effects, especially at high dose.

Another consequence of the fact that the overall RBE is controlled by RBE_{max} and RBE_{min}, and that these parameters are, respectively, inversely and directly related to the reference radiation α/β ratio or its square root (see appendix of this chapter), then a practical RBE for a clinically used dose cannot be assumed to be only related to RBE_{max}. Mathematical models that predict the RBE from the inverse or reciprocal value of α/β must be incomplete. A better approach, shown in chapters 8 and 9, enables separate modification of α and β with increasing LET, and consequently allows the RBE to be related to the reciprocal of α/β at low dose,

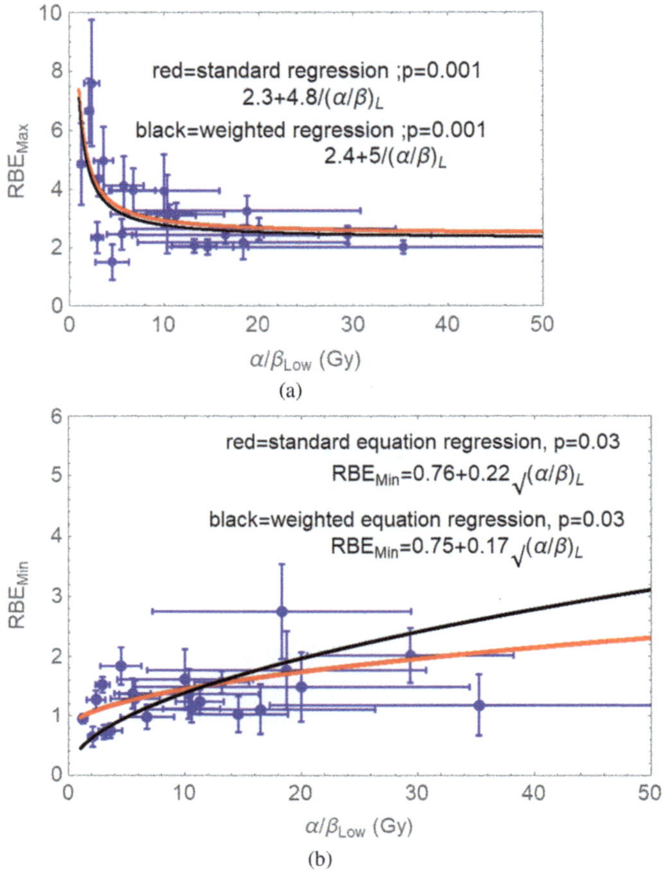

Figure 5.1. Plots of RBE_{max} (a) and RBE_{min} (b) against the reference low LET α/β ratio, adapted from Jones *et al* (2011).

but to be dependent on $\sqrt{(\alpha/\beta)}$ at high dose, with a gradual transition between these two limits with change of dose.

Warenius and colleagues (Warenius and Britten 1994, Warenius *et al* 1994) identified that the respective rank order of radiosensitivities for photons and neutrons were not the same. This is a consequence of the above transition between RBE_{max} and RBE_{min} and other effects such as the saturation or reduction of RBE increments with increasing intrinsic (photon) radiosensitivity. This is discussed further in chapter 9 with simulation of what might be expected for protons.

It has also been shown that the LET values produced by fast neutrons increases the α-radiosensitivity parameter by a factor of 3.17, but the β parameter increases as $1.59\sqrt{\beta}$ on average. Previously, it was taught that β did not increase with LET (as evidenced in meticulous experiments involving only one cell line, the V-79 cell line, by Chapman (2003)), but the Clatterbridge *in vitro* neutron data set contained over twenty cell lines and the increase in β can be seen in figure 5.3 (Jones 2010).

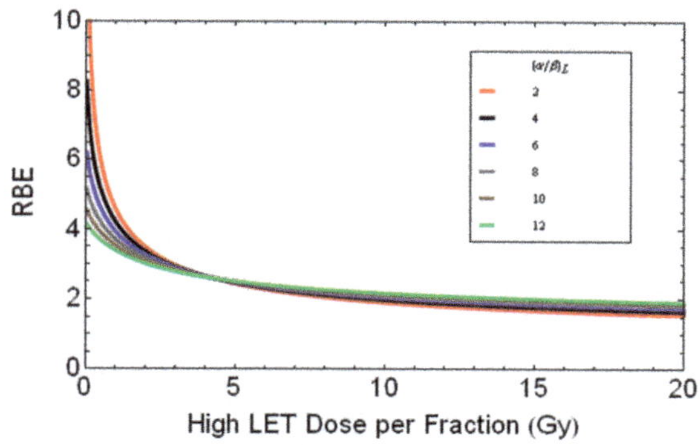

(a)

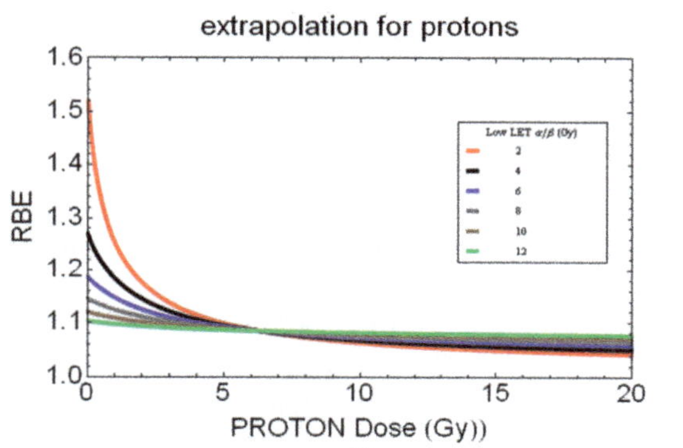

(b)

Figure 5.2. Relationship between (a) fast neutron dose per fraction and the RBE for different α/β ratios, and (b) the transformation of the above to provide a near flat response for α/β of 10 Gy to simulate proton data, as discussed in chapters 7 and 9. The red curves suggest how brain and spinal cord tissues may behave ($\alpha/\beta = 2$ Gy), followed by a gradual change in α/β to faster growing systems such as many rapidly growing tumour types and acute reacting normal tissues ($\alpha/\beta = 10$ Gy). Reproduced from Jones (2011).

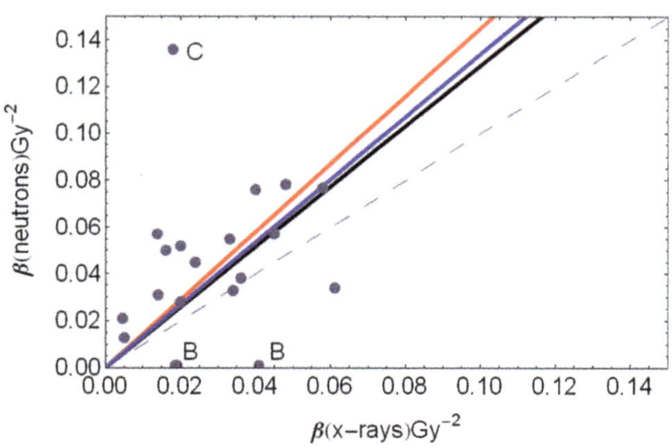

Figure 5.3. Cell lines labelled C and B, B were excluded due to known radiation repair deficiencies and extreme values, β being difficult to measure. The hatched line is the expected result for no change in β with increasing LET. Most cell line data points are above that line. The V-79 line falls just above it (the seventh point from the right-hand side). Reproduced from Jones *et al* (2011).

There now remain very few advocates for neutron therapy in the world. In fact the only promising development for neutrons is arguably boron neutron capture therapy—a complex binary therapy that involves a low dose exposure to thermal or epithermal neutrons, which are selectively captured by boron-labelled molecules taken up by rapidly growing tumour cells. The reaction creates more intense localized ionization by generating an alpha particle and a lithium ion, with tissue ranges of only around 10 μM (around one cell diameter). Some pilot studies have shown promise in highly malignant brain tumours, or for recurrent tumours, although in uncontrolled, highly selected patients (Kageji *et al* 2014, Lim *et al* 2015). Further discussion is beyond the scope of this book.

Following the clinical fast neutron trials, the UK funding authorities decided not to invest in further ambitious radiation projects, and as a result a general decline in radiotherapy research followed, although the Clatterbridge cyclotron was successfully adapted for proton therapy in the eye. There emerged a clinical scepticism regarding the use of cyclotrons in radiotherapy and high LET radiations in general. In other countries, more progress was achieved using charged particle therapy (protons or light ions) produced from cyclotrons or the larger synchrotrons. Some of these countries had no prior experience of high LET radiations in the form of neutrons.

A major radiotherapy research question over the next few decades will be whether or not carbon ions are superior to protons in specific cancers, because of better beam ballistics and dose localization and/or the apparent greater reduction in the oxygen effect. Suit *et al* (2010) pointed out that since even the neutron studies did not show a convincing overall local tumour control improvement, it should not be expected that carbon ions will provide improved efficacy since the oxygen effect dependency is greater for spread out carbon ion peaks than for fast neutrons (OER ratios of around 1.8–2 compared with 1.6–1.8, respectively).

The process of reoxygenation in tumours during radiotherapy, lasting a week or more, may be the reason why hypoxic radioresistance is not especially important; dose escalation for hypofractionation may also overcome initial hypoxia, but it is intriguing that identification of tumours with slow reoxygenation kinetics may yet be important since high LET therapy and dose escalation may be important. Again the reality of treating tumours *in situ* appears to be more difficult than the *in vitro* treatment of cells for many reasons.

5.5 Some important conclusions

- There can be no complacency about x-ray radiotherapy, as some patients continue to develop severe, chronic and debilitating side effects, so that prescribed doses are often limited by clinical decisions taken regarding the probability of such complications in each individual.
- It is important that the same mistakes made with neutrons are not repeated with charged particles.
- Radiobiological testing should be sufficiently comprehensive, including many cell lines of various radiobiological characteristics, and not just a few or one.

- In retrospect, better radiobiological modelling would have alerted clinicians of at least some of the adverse features.
- All new forms of radiotherapy need to be tested in high-quality centres, using the best input from physics, biology and medicine.
- The experimental fast neutron database remains important and has implications for proton therapy because neutrons mainly ionize by forming recoil protons.

Appendix

RBE_{max} and RBE_{min} act as multipliers of the low LET α and β parameters. By considering low and high dose limits, these parameters have the following identities (the subscripts L and H refer to low and high LET radiations respectively).

$$RBE_{max} = \frac{\alpha_H}{\alpha_L} \tag{5.1}$$

so that

$$\alpha_L = \frac{\alpha_H}{R_{max}} \tag{5.2}$$

and

$$RBE_{min}^2 = \frac{\beta_H}{\beta_L} \tag{5.3}$$

so that

$$\beta_L = \frac{\beta_H}{RBE_{min}^2}. \tag{5.4}$$

If equation (5.2) is divided by equation (5.4), then

$$\frac{\alpha_L}{\beta_L} = \frac{\alpha_H}{\beta_H}\frac{RBE_{min}^2}{RBE_{max}}. \tag{5.5}$$

Rearrangement of this last equation allow RBE_{max} and RBE_{min} to be expressed as:

$$RBE_{max} = \frac{\frac{\alpha_H}{\beta_H} RBE_{min}^2}{\left(\frac{\alpha_L}{\beta_L}\right)} = \frac{Q}{\left(\frac{\alpha_L}{\beta_L}\right)} \tag{5.6}$$

and

$$RBE_{min} = \sqrt{\frac{\frac{\alpha_L}{\beta_L} RBE_{max}}{\frac{\alpha_H}{\beta_H}}} = S\sqrt{\frac{\alpha_L}{\beta_L}} \tag{5.7}$$

where

$$Q = \frac{\alpha_{\mathrm{H}}}{\beta_{\mathrm{H}}}\mathrm{RBE}_{\min}^2 \tag{5.8}$$

and

$$S = \sqrt{\frac{\mathrm{RBE}_{\max}}{\frac{\alpha_{\mathrm{H}}}{\beta_{\mathrm{H}}}}}. \tag{5.9}$$

References

Bewley D K 1989 *The Physics and Radiobiology of Fast Neutron Beams* (Bristol: Institute of Physics)

Carabe-Fernandez A, Dale R G, Hopewell J W, Jones B and Paganetti H 2010 Fractionation effects in particle radiotherapy: implications for hypo-fractionation regimes *Phys. Med. Biol.* **55** 5685–700

Carabe-Fernandez A, Dale R G and Jones B 2007 The incorporation of the concept of minimum RBE ($\mathrm{RBE}_{\min}$) into the linear-quadratic model and the potential for improved radiobiological analysis of high-LET treatments *Int. J. Radiat. Biol.* **83** 27–39

Chapman D J 2003 Single-hit mechanism of tumour cell killing by radiation *Int. J. Radiat. Biol.* **79** 71–81

Dale R G and Jones B 1998 The assessment of RBE effects using the concept of biologically effective dose *Int. J. Radiat. Oncol. Biol. Phys.* **43** 639–45

Dearnaley D P *et al* 2014 Escalated-dose versus control-dose conformal radiotherapy for prostate cancer: long-term results from the MRC RT01 randomised controlled trial *Lancet Oncol.* **15** 464–73

Duncan W 1994 An evaluation of the results of neutron therapy trials *Acta Oncol.* **33** 299–306

Field S B 1976 An historical survey of radiobiology and radiotherapy with fast neutrons *Curr. Top. Radiat. Res. Q.* **11** 1–86

Gray L H, Mottram J C and Read J 1940 Some experiments upon the biological effects of fast neutrons *Brit. J. Radiol.* **13** 371–88

Gray L H, Conger A D, Ebert M, Hornsey S and Scott O C A 1953 The concentration of oxygen dissolved in tissues at the time of irradiation as a factor in radiotherapy *Brit. J. Radiol.* **26** 638–48

Henk J M, Kunkler P B and Smith C W 1977 Radiotherapy and hyperbaric oxygen in head and neck cancer. Final report of first controlled clinical trial *Lancet* **16** 101–3

Jones B 2010 The apparent increase in the β-parameter of the linear quadratic model with increased linear energy transfer during particle irradiation *Brit. J. Radiol.* **83** 433–6

Jones B and Hendry J 2017 Professor Jack Fowler and Oliver Scott *Brit. J. Radiol.* **90** 20160904

Jones B, Underwood T C, Carabe-Fernandez A and Dale R G 2011 Further analysis of fast neutron relative biological effects and implications for charged particle therapy *Brit. J. Radiol.* **84** S11–1

Kageji T, Nagahiro S, Mizobuchi Y, Matsuzaki K, Nakagawa Y and Kumada H 2014 Boron neutron capture therapy (BNCT) for newly-diagnosed glioblastoma: comparison of clinical results obtained with BNCT and conventional treatment *J. Med. Invest.* **61** 254–63

Laramore G E *et al* 1993 Neutron versus photon irradiation for un-resectable salivary gland tumors: final report of an RTOG-MRC randomized clinical trial *Int. J. Radiat. Oncol. Biol. Phys.* **27** 235–40

Lim D, Quah D S, Leech M and Marignol L 2015 Clinical potential of boron neutron capture therapy for locally recurrent inoperable previously irradiated head and neck cancer *Appl. Radiat. Isot.* **106** 237–41

Nutting C M *et al* 2011 Parotid-sparing intensity modulated versus conventional radiotherapy in head and neck cancer (PARSPORT): a phase 3 multicentre randomised controlled trial *Lancet Oncol.* **12** 127–36

Suit H *et al* 2010 Proton vs carbon ion beams in the definitive radiation treatment of cancer patients *Radiother. Oncol.* **95** 3–22

Underwood T and Paganetti H 2016 Variable proton relative biological effectiveness: how do we move forward? *Int. J. Radiat. Oncol. Biol. Phys.* **95** 56–8

Underwood 2017 Private communication

Warenius H M, Britten R A, Browning P G, Morton I E and Peacock J H 1994 Identification of human *in vitro* cell lines with greater intrinsic cellular radiosensitivity to 62.5 MeV (p-->Be+) neutrons than 4 MeV photons *Int. J. Radiat. Oncol. Biol. Phys.* **28** 913–20

Warenius H M and Britten R A 1994 *In vitro* studies of intrinsic cellular radiosensitivity following 4 MeV photons or 62.5 MeV (p-->Be+) neutrons. Potential implications for high LET therapy *Acta Oncol.* **33** 241–9

Chapter 6

Fractionation

6.1 Introduction and background radiobiology

The term 'fractionation' refers to the splitting of dose into separate treatment 'fractions', or treatment sessions; it is an important part of the treatment prescription. The splitting process carries further consequential implications on the overall treatment time, depending on the inter-fraction time interval used. Fractionation is important because it can be used to either improve clinical outcomes, or alternatively to achieve closely similar outcomes at reduced financial cost. For example, a greater degree of fractionation of the same total dose reduces most late normal tissue side effects, whereas the same biological endpoint can be achieved by two very different fractionation schedules provided the total dose is changed to accommodate the degree of fractionation used. Time is a further parameter within the radiotherapy prescription, since the overall treatment duration is related to the number of fractions multiplied by the average inter-fraction interval. Increasing the time normally reduces acute normal tissue reactions but the use of too closely spaced fractions can enhance all normal tissue effects. Fractionation is often referred to as 'time dose fractionation'.

There is a large literature on fractionation of conventional x-ray (photon)-based therapy, which varies from relatively simple qualitative considerations to more complex mathematics. The process of fractionation includes key variables, which are in most situations under medical control: the dose per treatment (d), the number of treatments (n), the total dose D (or $n \times d$) and the overall treatment time (t). In general, when n is increased, d decreases (to an extent that D almost always increases) in order to preserve the same bio-effect. This occurs due to sub-lethal radiation damage repair between fractions; the greater the fractionation the greater the opportunity for repair between fractions. It follows that cells that have defects in damage recognition or repair pathways, as found in mutated tumour cells, will be more radiosensitive and less fraction sensitive. Prolongation of t may also require increases in d or n, or both, in order to preserve a biological iso-effect such as to

doi:10.1088/978-0-7503-1338-4ch6

match tumour cell repopulation with time in the case of tumours that contain rapidly dividing cells, or normal epithelial tissues that contain rapidly growing cell populations. The relative magnitudes of these changes vary in different normal tissues and tumours, depending on their biological properties.

- Slow proliferation states, associated with a high repair capacity, have high fractionation sensitivity; that is a large change in effect follows a small change in d. These conditions are characterised by a low α/β ratio.
- Fast proliferation states, associated with a lower repair capacity, have a lower fractionation sensitivity; that is a much smaller change in effect follows a small change in d. These conditions are characterised by a high α/β ratio.

Although there have been many attempts to characterise fractionation 'rules' in terms of equations, the most successful and widely used model is that of the linear-quadratic (LQ) model of radiation effect, which contains α and β radiosensitivity parameters and for fractionation studies the α/β ratio is important since it is essentially inversely related to fractionation sensitivity, although the terms are sometimes used synonymously.

The term hypofractionation may require clarification. This essentially refers to a shift in fractionation to larger doses per fraction than 2 Gy by using fewer numbers of fractions. The converse is hyperfractionation. These terms can be confusing to some readers. Accelerated fractionation refers to dose schedules that deliver greater than 2 Gy (of conventional radiotherapy photons or x-rays) per day, independently of the number of fractions per day or week.

6.2 A brief history of fractionation

6.2.1 Radiobiology

Fowler (1984) summarised the extensive work on fractionation experiments and their analysis, emphasising the dominance of the dose per fraction effects on total dose requirements when compared with overall time for experimental acute skin reactions; as the dose per fraction reduces, a greater number of fractions is required for greater amounts of repair between these fractions. Time related effects are related to cellular repopulation, which is a slower process. He had also applied LQ theory, originally an empirical description in the 1940s of the yield of lethal chromosomal events with dose per cell by Lea (1962), to the study of iso-effective doses (Douglas and Fowler 1976) using an algebraic reconfiguration of the LQ model in the form of an FE plot (fractionation effect).

The linear-quadratic effect, E (or yield of lethal chromosomal aberrations in terms of Lea's findings), is expressed in fractionated form as:

$$E = n(\alpha d + \beta d^2) \tag{6.1}$$

It follows, if $D = nd$, that

$$\frac{1}{D} = \frac{\alpha}{E} + \frac{\beta}{E}d. \tag{6.2}$$

This transformation to linearity allowed Douglas and Fowler (1976) to study iso-effective doses in their FE plots of *1/D* against *d*. By taking the ratio of the intercept to the slope they showed that α/β ratios differed for acute and late effects. They also, importantly, showed that α/β values were far higher for fast neutron fractionation experiments, including in pig skin, which closely resembles human skin in radiation tolerance. This finding correlated with the fact that increased fractionation does not require as great a change in total dose as in the case of x-rays (photons) to preserve an iso-effect.

The time effect, measured as the additional radiation dose required to maintain the same degree of skin reaction, progressively increased after a finite time delay; this was due to repopulation of skin commencing only after radiation damage expression, which leads to the concept of accelerated repopulation in early reacting tissues (Denekamp 1973). It is therefore inappropriate to use a power law model for such an effect since the change in iso-effective dose with time requires a variable slope, even on a logarithmic plot. For tumour control, Fowler *et al* (1974) found an optimum overall time in an experimental tumour while maintaining a normal tissue (skin) iso-effect. This achievement was attributed to effective tumour reoxygenation at the optimum overall time since equally high tumour control was found at much shorter overall times either by the addition of metronidazole (as a hypoxic cell sensitiser), or by using fast neutrons (due to their reduced dependency on oxygen for cell killing, often referred to as a reduced oxygen enhancement ratio). At longer overall times than the optimum time, tumour control was progressively reduced, probably due to tumour cell repopulation.

During the 1980s further progress included the detailed analysis of Withers that acute reacting tissues (by then characterised by high α/β values) required little change in total dose with increasing fractionation, whereas late reacting tissues (with low α/β ratios) required more substantial changes in total dose; the form of the relationship, with an asymptotic limit at very small values of d (and large n) was compatible with LQ theory. Towards the end of the decade several advances occurred, including:

1. Inclusion of a repopulation factor alongside the LQ surviving fraction equation was an important advance (by Dale (1989) and Fowler (1989)), which allowed different fractionation schedules to be compared. Fowler could then argue that clinical outcomes with quite different fractionation schedules could be broadly equivalent.
2. Evolution of the LQ model to the biological effective dose (BED) concept following the original idea of Barendsen and subsequent important adaptations by Dale and Fowler. This allowed calculation of iso-effective (or equi-effective) treatment schedules, ranking of different fractionation schedules, addition of multiple phases of treatment by different radiation modalities or different fractionation patterns, different dose rates (Jones *et al* 2001, Fowler 2010) and eventually the inclusion of relative biological effect (RBE) limits, as originally described by Bewley, in BED equations for high linear energy transfer (LET) radiations (Jones *et al* 2006, Carabe-Fernandez *et al* 2007), and which allows a variable or flexible RBE to change with dose per fraction, as will be shown later.

6.2.2 A synopsis of clinical fractionation

Accounts of the history of fractionation are available in Thames and Hendry (1987), with a good description of the failure of power law fractionation models in the clinic.

Progress in clinical fractionation was made by careful empirical practice and observation. Following the lead of Regaud in France, who used very low dose rate radium applications, Coutard noted that protraction and hyperfractionation of teletherapy to high total doses produced a substantial reduction in both acute and late effects of therapy while maintaining the desired tumour effect.

The advent of megavoltage radiotherapy, with improved depth dosage, led to further and more formal studies, such as the studies at The Hammersmith Hospital. There, patients were treated with several total doses, but the overall time remained constant since the number of fractions was fixed at twenty. The results effectively confirmed that small increments in dose per fraction and total dose caused both enhanced toxicity and tumour control. A general trend emerged: cancer centres in the North of England, Scotland and Canada favoured daily hypofractionated radiotherapy at 2.5–3.3 Gy per fraction in 15–20 fractions over 3–4 weeks, while those in the South of England favoured the French/American approach of conventional daily doses of 1.8–2 Gy given in 30–35 fractions (five fractions per week) over 6–7 weeks.

The concept of hyperfractionation combined with acceleration emerged in the early 1980s and followed advances in the understanding of the radiobiology of cell killing, repair processes and repopulation effects.

Because of persistent economic constraints in the provision of radiotherapy facilities, there remains considerable interest in the use of relatively hypofractionated radiotherapy, especially for delivery of sub-radical doses in an adjuvant setting, such as pre-operative rectal cancer or after wide excision of a primary breast cancer. The UK START breast cancer trial was designed to test daily 2 Gy fractions against larger fraction sizes over the same overall treatment time and which showed broadly similar effects (Haviland *et al* 2013).

The importance of fractionation in radiotherapy cannot be doubted, because of its clinical impact. Increasing the number of fractions n, along with a reduction in d, inevitably increases t if, say, only a maximum of five fractions a week are to be given; this will allow acute reacting normal tissues (such as epithelial surfaces, skin, oesophagus, intestine) to tolerate radiation treatment better, since the increase in elapsed time will allow more cellular repopulation to occur. Exactly the same will occur with particle therapy, so that for large treatment volumes containing epithelial surfaces that can repopulate within a month, longer treatment courses tend to be given in order to spare acute effects. The opposite, over intensification of dose in time, can cause very severe acute effects that can lead to so-called 'consequential' late effects in tissues, as well as being very unpleasant to the cancer patient.

The situation with late tissue effects is different. In their case, increasing fraction size will have a more deleterious effect for x-rays (photons) than for positively charged particles; the former will require a larger reduction in total dose to maintain an iso-effect, whereas the latter will not require as great a change in total dose and so can be given in fewer fractions.

Late effects and very slow growing tumours generally require no time correction factor, unless treatment is very protracted or if multiple fractions per day (MFD) are used. For MFD, incomplete repair (IR) equations are required, especially for three fractions or more per day. IR is likely to be reduced for high LET radiations compared with low LET radiations, since IR is included within the LQ model by linkage to the β parameter and not the α parameter, which increases to a greater extent than β with LET, as explained previously. Another issue with IR, is fraction duration (t_f); difficult to deliver techniques allow short but significant interruptions of treatment, (due to gantry rotation, re-imaging, etc). This also applies to some modern photon techniques such as GammaKnife and linear accelerator treatments such as CyberKnife. This is because there are principally two repair half times: the best estimates remain those in rat spinal cord, namely the first component is fast (half time = 0.19 h) and the second slower (half time = 2.16 h) with close to 50% of damage associated with either form of repair (Pop *et al* 2000). It has been shown that many treatments with prolonged t_f values can result in a significant reduction in BED per fraction (Hopewell *et al* 2012). It can be argued that all fractional durations longer than 30 min should be recorded for later analysis. Ion beam treatments such as the large 40–44 Gy-eq single fractions given for early stage lung cancer in older Japanese patients can in some instances take over 2 h to deliver (Karube *et al* 2016). The intervals of time within a fraction that allow repair to continue between 'pulses' of treatment will affect low LET therapy to a greater extent than high LET. Also, tumours and tissues with a high α/β are less sensitive to dose rate and incomplete repair effects than their low α/β counterparts where extension of t_f produces a greater reduction in BED. More detailed work is required on this topic in the context of hadrontherapy, using prospective data recorded in the ion beam centres to be eventually correlated with clinical outcomes.

It must always be remembered that a small increment in dose per fraction beyond that prescribed, if deposited in normal tissue, can lead to enhanced tissue damage due to the nonlinear effects associated with the LQ model. This has been termed double trouble at 2 Gy per fraction and this effect is enhanced when the dose per fraction increases further beyond 2 Gy in hypofractionated radiotherapy, when it is called treble (or triple) trouble (Jones *et al* 2000, Jones *et al* 2001). It follows that increasing hypofractionation (larger doses per fraction) must be given with caution with respect to dose homogeneity and individual calculations should be done since RBE effects will also complicate this effect, as shown later in this book.

Much could be written here on the economic advantages of hypofractionation; fewer treatments of larger doses per fraction allows greater throughput of patients per year, reduced individual patient costs and inconvenience, with greater utility within society bearing in mind strong competition for finite health care resources. Such potential advantages can only be pursued if outcomes, in terms of tumours controlled and cured along with satisfactory quality of life statistics, either match expectations or can be proven to be superior to other more expensive approaches. The use of adjuvant therapies, either conventional or experimental, must also be considered with fractionation trials, since they may overcome certain disadvantages (e.g. cetuximab on tumour repopulation for longer schedules, new antimetabolites which reduce oxygen consumption and permit rapid tumour reoxygenation prior to short schedules).

Also, it remains to be shown if protons can be delivered in hypofractionated form as safely as carbon ions. It is paradoxical that the majority of patients treated with protons had eye melanomas, treated in four large fractions to 52–54 Gy doses, using 'small fields'. The large and protective volume effect will tend to favour larger fractions for small target volumes as long as non-essential tissues are included in the tumour–normal tissue margin. Our inadequate, but improving, knowledge of the mathematical relationship between normal tissue volume and radio-tolerance may yet provide safer guidelines in the future.

Since most randomised fractionation trials have shown only marginal differences, it is likely that higher gains will only be achieved by treatment individualisation, or at least treatment determination according to an analytical formula based on predictive assays and other information. Random sampling simulations of virtual trials, using radiobiological modelling where treatment is allocated according to radiobiological characteristics such as radio-sensitivity and repopulation, produce considerably greater gains in tumour control probabilities while respecting normal tissue tolerance (Jones and Dale 2000); this represents personalisation of treatment and the scope for this approach is perhaps even greater in the case of particle therapy.

6.3 Modelling of fractionation

6.3.1 LQ modelling of fractionation in high LET radiations with inclusion of the RBE

In comparison with conventional x-ray based fractionation experience, detailed fractionation experiments using high LET radiation have only a 50 year history (for example, Barendsen and Broerse 1970, Fowler *et al* 1976). They show reduced dependency on fractionation effects compared with low LET radiations:

$$\text{RBE} = \frac{d_\text{L}}{d_\text{H}}, \tag{6.3}$$

where the subscripts L and H refer to low and high LET respectively.

Now, when the fractionation schedule changes, the numerator d_L changes with fractionation to a greater extent than d_H. The low LET numerator dose changes because of repair occurring between fractions, but the denominator (d_H) is less susceptible to changes in fractionation. This is because high LET radiations form more clustered, locally complex damage within DNA, which is more difficult or impossible to repair. So, there is less repair between fractions in the case of high LET radiations compared with low LET radiations. This must also be understood in terms of the single fraction experiment, where there is no opportunity for repair between fractions. With increasing LET, the increases in the α radiosensitivity parameter exceed those of the β parameter; this means that the ratio of $\alpha_\text{H}/\alpha_\text{L}$ should always exceed $\beta_\text{H}/\beta_\text{L}$. Now since $\alpha_\text{H}/\alpha_\text{L}$ is RBE_max (the RBE at near zero dose where β related cell kill is negligible) and $\sqrt{(\beta_\text{H}/\beta_\text{L})}$ is RBE_min (the RBE at high dose where β related cell kill predominates). It follows that a reduction in dose per fraction will increase the RBE towards the RBE_{max} limit; conversely as the dose increases the RBE will approach the lower RBE_{min} value. It can also be appreciated that the yield of complex lethal chromosomal aberrations per unit dose increases with LET, so the same biological end points can be achieved with lower doses, where the α cell killing mechanism predominates, rather than the β related mechanism. This again means that radiations that have high LET and RBEs will be associated with reduced fractionation sensitivity, since it is the β related damage that is susceptible to sub-lethal damage repair.

This original description of the RBE was convenient to radiobiologists, since the RBE could be measured from the defining ratio given in equation (2.4), but this allowed the RBE to be viewed as being over simplistic, indeed almost of trivial complexity. In fact the RBE is a highly complex parameter, depending on multiple variables including particle mass, energy, charge, target volume (and its depth), dose, beam contamination (e.g. with neutrons, γ-rays, etc) and the radiobiological characteristics of the tissue/tumour being targeted (repair capacity, proliferative status—linked to the low LET α/β).

For simplification, reduction of type-space and to obtain greater clarity for understanding the radiobiological principles, the text below contains RBE parameters without reference to specific LET values. Some of the methods for estimating RBE_{max} and RBE_{min} from LET are given separately in chapters 9 and 10, and can be used to provide these RBE parameters for individual calculations.

6.3.2 BED equations

The BED equations that include the RBE have already been given in chapter 2. They will now be used to illustrate fractionation effects

The BED equation can be modified for hadrontherapy (Jones *et al* 2006, Carabe-Fernandez *et al* 2007), has already been derived in chapter 2, and is reproduced here as

$$\mathrm{BED} = D_{\mathrm{H}}\left(\mathrm{RBE}_{\max} + \frac{\mathrm{RBE}_{\min}^{2}\, d_{\mathrm{H}}}{\alpha/\beta}\right). \tag{6.4}$$

The basic BED principle can also be extended to provide EQD_2 equations, which express the dose required for a defined biological or clinical endpoint when the dose is given in 2 Gy fractions; EQD_2 is simply the BED divided by $(1 + 2/(\alpha/\beta))$.

The function of the α/β ratio is to act as a coefficient that controls the fractionation sensitivity. When fraction size is changed, the resulting total dose needs to be modified in order to provide a constant BED (and the same bio-effect). It is important to understand that different tissues and tumours can possess quite different α/β ratios, so that they respond differently to different changes in dose fractionation (for further details see chapter 2).

When the two limits of RBE are used, the α/β ratio must be the same as that for the low LET (x-ray) case, and this parameter will now be less important in influencing the BED when d_H is changed, compared to the standard low LET equation, especially when RBE_{max} is large. The percentage BED increment with dose per fraction is consequently larger for low LET treatments compared to higher LETs.

This equation, when used to equate with a low LET schedule, allows the RBE at any intermediate dose to vary between the RBE_{max} and RBE_{min} values used.

6.3.3 Converting a specific low LET BED fractionation to that for high LET, when the low LET α/β ratio is known, but with no change in overall treatment time

It is then possible to pursue iso-effect calculations using the equality:

$$\text{low LET BED} = \text{high LET BED}$$

$$D_L\left(1+\frac{d_L}{\left(\frac{\alpha}{\beta}\right)_L}\right) = D_H\left(RBE_{max} + \frac{RBE_{min}{}^2\, d_H}{\left(\frac{\alpha}{\beta}\right)_L}\right). \tag{6.5}$$

The major advantage here is that the α/β ratio used throughout this equality is that of the low LET α/β, which is well established for some tissue types, e.g. brain and spinal cord, although it is necessary to know the likely values of RBE_{max} and RBE_{min}. In important clinical scenarios these RBE parameters should be sufficiently large to produce conservative values of dose per fraction for the protection of normal tissues with low α/β values, yet sufficiently low in the case of fast growing tumours with large α/β ratios.

Worked example

It is decided to design a clinical trial that provides the same BED for late brain effects as 60 Gy in 30 fractions of megavoltage x-rays, using either 10 or 15 fractions of carbon ions, which are assumed to have RBE_{max} values of 6 and RBE_{min} of 1.25. Find the dose required if given in 2, 3 or 4 Gy fractions. The brain α/β is 2 Gy. Compare the results with an assumption that the dose is modified by constant RBE factors of (a) 3 and (b) 4, regardless of the magnitude of dose. (*Caveat: the parameters and numbers used here are chosen for the convenience of those using pocket calculators*).

The BED for x-rays is:

$$60(1 + 2/2) = 120\ Gy_{[2]}$$

and this BED must now equate with the high LET BED such that $120 = nd_H(RBE_{max} + RBE_{min}{}^2 d_H/k)$, where k is from now on the low LET α/β ratio.

So, for $n = 10$, $d_H = 1.65$ Gy and $n = 15$, $d_H = 1.16$ Gy.

If doses (d_H) of 2 Gy, 3Gy or 4 Gy fractions are to be given the solutions to the above equation provides $n = 7.9$, 4.79 and 3.29. These are not close to integer values. Rounding off to the nearest fraction would be inappropriate, resulting in under- or over-dosage. Consequently, it is best to use the former approach, which calculates the dose per fraction.

Now if the 'overall RBE' is a constant at all doses, it has a purely dose modifying function, so that each d term in the BED equation is multiplied by the RBE, then we must solve for d_H in:

$$120 = n\, d_H(\mathrm{RBE} + \mathrm{RBE}^2 d_H/k)\ (\text{or } 120 = n\, d_H \mathrm{RBE}(1 + \mathrm{RBE} d_H/k),$$

so that for RBE = 3, $n = 10$, $d_H = 1.33$ Gy and $n = 15$, $d_H = 1.04$ Gy, and for RBE = 4, $n = 10$, $d_H = 1.00$ Gy and $n = 15$, $d_H = 0.78$ Gy.

These results show the critical dependency of the RBE model (fixed versus flexible) for the assumptions used.

6.3.4 Overall fractionation differences between low and high LET radiations

Solution of the iso-effect equations given above (equation 3.6 is solved for d_L and the result divided by d_H to provide the RBE at any value of d_H) can be used to show how low LET (figure 6.1) and high LET (figure 6.2) fractionation influences the total dose required to maintain a tissue iso-effect. The change in total dose with number of fractions (and consequently dose per fraction) is much greater for the low LET megavoltage x-rays. The dependency on tissue α/β can also be seen, with a pronounced flattening of the relationship after a few fractions only for the high LET radiation; the change over the first few fractions indicates the shift from reliance on RBE_{min} for the first few fractions to a greater dependency on RBE_{max} after that. There appears to be little change in total dose after the 4–6 fraction schedules, with only marginal changes after 10–20 fractions. These modelled examples are typical of the pioneering fast neutron experiments (Fowler 1981). Similar proton beam iso-effects are given in chapter 9.

In PCP therapy, it is important to remember that these considerations apply to the high LET part of the beam; fractionation effects corresponding more closely to the low LET will be found where there are only entry beams that are distant from the Bragg peak regions. Thus, if, say, the skin and rib dose (mainly low LET) determine tolerance for a lung cancer treatment, as is the case in Japan, then fractionation will continue to have an important function. The incidence of rib fractures with large single doses remains of concern and this would probably fall with modest further

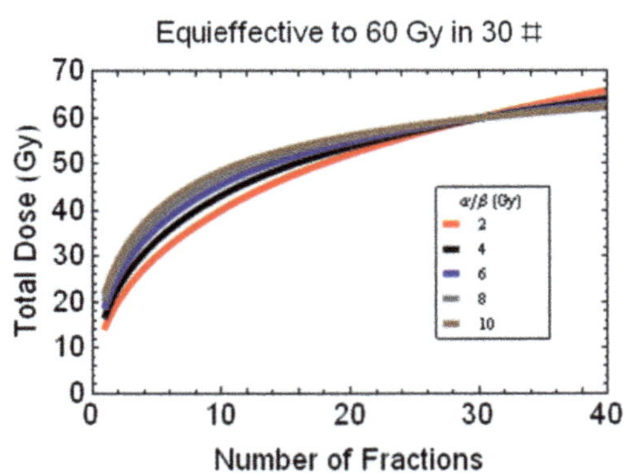

Figure 6.1. Total dose and number of fractions (of different fraction sizes) for megavoltage x-rays.

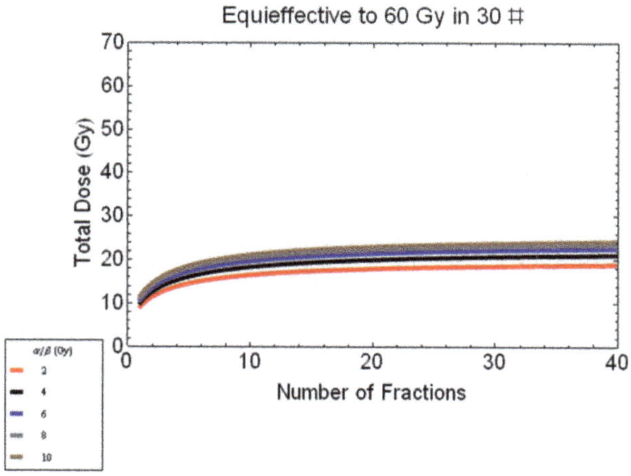

Figure 6.2. Total dose and number of fractions (of different fraction sizes) for carbon ions or fast neutrons assuming $RBE_{max} = 6$ and $RBE_{min} = 1.25$.

fractionation since the ribs are not in the very high LET region. 3D LET, dose and BED plans will ideally be required. LET-based treatment plans have already been generated by Grassburger *et al* (2011), although in the head region.

6.3.5 Boost doses

Radiotherapy is often the case where treatment is given in phases, typically beginning with wider field arrangements, often to cover regional lymphatic tissues, but ending with a boost dose to a smaller volume (the tumour itself with appropriate infiltrative margins); with x-ray based IMRT the boost can be given simultaneously with the wider field. Sometimes different radiation modalities are used for each phase: for example the larger initial treatment could be given by megavoltage IMRT, followed by the final phase using either brachytherapy, electrons, protons or ionic beams.

Worked example
A dose of 45 Gy in 25 fractions delivered by IMRT is followed by a carbon ion boost, with the aim of giving a total tumour dose equivalent to 70 Gy in 35 fractions of megavoltage x-rays, but preserving a normal tissue iso-effect, equivalent to 40 Gy in 20 fractions. Find the dose per fraction required for a 3, 4 or 5 fraction boost for the same tumour control, but also give the normal tissue BED values for each case. The likely tumour α/β is 4 Gy with $RBE_{max} = 4$ and $RBE_{min} = 1.4$; the critical late normal tissue has α/β of 3 Gy and $RBE_{max} = 4.7$ and $RBE_{min} = 1.3$ and the most relevant normal tissue volume receives a modal 72% of the IMRT dose and only 42% of the ion beam dose.

The intended total tumour BED is 70 (1 + 2/4) = 105 $Gy_{[4]}$.

The IMRT will provide a tumour BED of 45 (1 + 1.8/4) = 65.25 $Gy_{[4]}$.

The deficit tumour BED to be provided is 105 − 65.25 = 39.75 $Gy_{[4]}$.

The allowed total NT BED will be 40 (1 + 2/3) = 66.67 $Gy_{[3]}$.

The IMRT will provide a NT BED of 0.72 45 (1 + 0.72 1.8/3) = 46.40 $Gy_{[3]}$.

For tumour boost dose per fraction find d_H for n = 3, 4 and 5 in:

$39.75 = n\, d_H\, (RBE_{max} + RBE_{min}{}^2 d_H/4)$.

For n = 3, then d_H = 2.53 Gy; for n = 4, then d_H = 2.00 Gy and if n = 5, then d_H = 1.65 Gy.

The normal tissue BEDs will be, for each dose per fraction:

$0.72\ 45\ (1 + 0.72\ 1.8/3) + 3 \times 2.53 \times 0.42\ (4.7 + 1.3^2 \times 2.53 \times 0.42/3) = 62.87\ Gy_{[3]}$

$0.72\ 45\ (1 + 0.72\ 1.8/3) + 4 \times 2.00 \times 0.42\ (4.7 + 1.3^2 \times 2.00 \times 0.42/3) = 63.34\ Gy_{[3]}$

$0.72\ 45\ (1 + 0.72\ 1.8/3) + 5 \times 1.65 \times 0.42\ (4.7 + 1.3^2 \times 1.65 \times 0.42/3) = 63.74\ Gy_{[3]}$.

All of these BED values are below the allowed 66.67 $Gy_{[3]}$, but it is instructive to note that the lowest normal tissue BED is found for the least number of fractions (and the highest dose per fraction), because of the effective fall of RBE with hypofractionation in this case when using the extreme limits of RBE as parameters.

In other clinical situations, if normal tissue tolerance is exceeded, a BED calculation cannot provide the desired result for both normal tissue and tumour, because their BEDs are separately governed by different α/β and RBE parameters. A compromise in dose selection is necessary and this will require detailed clinical discussion, perhaps accepting increased risks of either recurrence or toxicity, or even a modest amount of each.

6.3.6 Converting a specific low LET BED fractionation to that for high LET, when the low LET α/β ratio is known, but with a change in overall treatment time

The equations have been used previously to estimate compensatory doses for unintended treatment interruptions.

$$\mathrm{BED} = D_L\left(1+\frac{d_L}{\left(\frac{\alpha}{\beta}\right)_L}\right) - KT_L = D_H\left(\mathrm{RBE}_{max} + \frac{\mathrm{RBE}_{min}{}^2\, d_H}{\left(\frac{\alpha}{\beta}\right)_L}\right) - KT_H,$$

and where K is the low LET BED dose equivalent of repopulation per day, K being defined as:

$$K = \frac{0.693}{\alpha_{L.}\omega},$$

where ω is the operative cellular doubling time. K values are known for some tumour types and are sometimes used along with a lag (or delay) time until the onset of accelerated repopulation. In such cases (usually assumed for squamous cell and transitional cell cancers) the T values are reduced by the delay time T_K, which is usually assumed to be 21–28 days. It is advantageous that the low LET K values can be used in this form, even within the high LET side of the equation, but provided the RBE factors are included within the BED portion of the equation.

6.3.7 Alternative approach for iso-effect calculations in the case of two high LET schedules

For two identical high LET treatments, fractionation iso-effects can be calculated using BED equations that contain high LET α/β ratios.

Thus for two iso-effective schedules of high LET we have for N_1 factions and N_2 fractions of dose d_1 and d_2

$$N_1H\left(\alpha_H d_1 H + \beta_H d_1^2 H\right) = N_2H\left(\alpha_H d_2 H + \beta_H d_2^2 H\right) \tag{6.6}$$

It is then permissible to divide throughout by α_H (and not α_L, as was previously the case) and obtain

$$D_{1H}\left(1+\frac{d_{1H}}{\left(\frac{\alpha}{\beta}\right)_H}\right) = D_{2H}\left(1+\frac{d_{2H}}{\left(\frac{\alpha}{\beta}\right)_H}\right). \tag{6.7}$$

In this case, the two RBE parameters RBE_{max} and RBE_{min} are no longer necessary. It would also be possible to use the single R_C conversion factor, where

$$R_C = RBE_{max}/RBE_{min}^2. \tag{6.8}$$

This is because

$$\left(\frac{\left(\frac{\alpha}{\beta}\right)_H}{\left(\frac{\alpha}{\beta}\right)_L}\right) = \left(\frac{\left(\frac{\alpha_H}{\alpha_L}\right)}{\left(\frac{\beta_H}{\beta_L}\right)}\right) = \frac{RBE_{max}}{RBE_{min}^2} \tag{6.9}$$

in the situation of comparing two identical high LET radiations, so that equation (6.7) would become

$$D_{1H}\left(1+\frac{d_{1H}}{R_C\left(\frac{\alpha}{\beta}\right)_L}\right) = D_{2H}\left(1+\frac{d_{2H}}{R_C\left(\frac{\alpha}{\beta}\right)_L}\right) \tag{6.10}$$

which could also be used in this form along with the time factor corrections, as given below.

For unintended treatment interruptions, the most appropriate time and repopulation correction factors can be added for the case of tumour iso-effects. This is not necessary for late reacting tissues. The equations are then:

$$D_{1H}\left(1+\frac{d_{1H}}{\left(\frac{\alpha}{\beta}\right)_H}\right) - K_H t_{1H} = D_{2H}\left(1+\frac{d_{2H}}{\left(\frac{\alpha}{\beta}\right)_H}\right) - K_H t_{2H} \tag{6.11}$$

where t_1 and t_2 are the respective overall treatment times for schedule 1 and 2 and K_H is defined as

$$K_H = \frac{0.693}{\alpha_{H.}\omega}. \tag{6.12}$$

Compared with the K values for low LET (K_L), there is little knowledge of K_H values in the literature. K_H can, however, be found from K_L by dividing by RBE_{max} (which is α_H/α_L), that is

$$K_H = \frac{K_L}{RBE_{max}}. \tag{6.13}$$

A note of caution must be given about using RBEs based on total fractionated dose, since this approach can lead to confusion and potential errors. It is best to use RBE on a dose per fraction basis, which is consistent with cell survival curve theory and the original definition of RBE. It is recommended that individual dose per fraction RBEs be calculated and then total dose calculated. Where fraction numbers differ between low and high LET schedules this is especially important.

For example, let us suppose that a 10 Gy carbon ion physical dose in five fractions produces the same late effect as a 50 Gy in a 25 fraction photon dose—what is the RBE?

By taking each of the:

- total doses, the RBE is 50/10 = 5;
- delivered doses per fraction, the RBE is 2/2, which is absurd.

However, by using the following procedure—based on corrected dose per fraction—a different answer is obtained.

The RBE denominator will be the high LET dose per fraction, which is 2 Gy. The photon dose per fraction, if given in five fractions (rather then 25), which would match the same effect as 50 Gy in 25 fractions of photons, assuming a late effect $\alpha/\beta = 3$ Gy, would be given by d in the following BED iso-effective doses:

BED (five fractions, unknown d) = BED (25 fractions, $d = 2$)

$$5d\left(1+\frac{d}{3}\right) = 50\left(1+\frac{2}{3}\right)$$

$d = 5.73$Gy, which is then used as the numerator of the RBE.

The RBE is then 5.73/2 = 2.87.

In situations where under- or over-dose has occurred, or in calculation of x-ray equivalent schedules, the most reasonable estimates of the parameters should be used; clinicians may prefer to use the safest choices, such as a high RBE_{max} in tissues that are highly fraction sensitive for x-rays, such as the central nervous system. The general advice for such calculations in the case of x-rays has been published elsewhere (e.g. Joiner 2004, Jones and Dale 2008), but the additional effect of the RBE must be included if there is any attempt to use (low LET) x-ray tissue tolerances or x-ray tumour control data. The alternative approach is to use only high LET α/β ratios within BED equations where the tolerance and tumour control BED values for the high LET radiation are known.

6.4 RBE and dose per fraction

Early RBE experiments, especially studies using fast neutrons, with RBE values similar to carbon ion therapy, showed that the RBE reduced with increasing dose per fraction, leading to potential clinical dangers for small fractions, or higher than expected bio-effects despite high LET dose fall off in normal tissues. Fall off of megavoltage x-ray dose causes a reduction in tissue dose per fraction, but fall off of high LET dose beyond a target volume may be, at least partially, compensated for by a higher RBE.

It is instructive to study figure 6.3, which shows the relationship between dose per fraction of neutrons and RBE for tissues with different α/β ratios; the overall change in RBE with dose per fraction is much less for biosystems with a high α/β than for those with a low α/β.

This was compiled using parameters derived from a variety of fast neutron experiments *in vivo* and on human cancer cells *in vitro* (Jones *et al* 2011), already shown in chapter 5. Due to RBE_{max} being inversely related to the low LET α/β, and RBE_{min} being linearly related to the square root of α/β, there is a crossing over of the RBE values such that biosystems with a low α/β have the highest RBEs at low doses, but the lowest RBE at high doses. If a cell or tissue system does not have a clearly identified α/β ratio then the choice of dose per fraction around the cross-over part of the curves would minimise the RBE uncertainty.

There are important implications for clinical fractionation here, although these predictions must be better validated by further experimental studies. It is interesting to note that lowering the dose per fraction probably caused more toxicity in late reacting tissues in UK neutron trials. Also, there are suggestions from the Japanese ion beam centres that late side effects may be less prevalent with hypofractionated schedules. Such claims must, however, be subjected to prospective studies, since at present the hypofractionated regimens have the shortest follow-up times and so may show bias. However, these findings are compatible with a hypothesis that the rank order of RBE values can be reversed according to the dose level, as in figure 6.3. In the case of the brain, with the lowest α/β of 2 Gy, this may be important, since the therapeutic ratio might be improved by hypofractionation as long as the brain RBE at large doses is sufficiently lower than the tumour RBE. This may already explain some excellent outcomes at Heidelberg and GSI for skull base tumours compared with low dose per fraction protons in USA, although the shorter treatment time (<20 days) may also be operative in at least some tumours, which may contain more

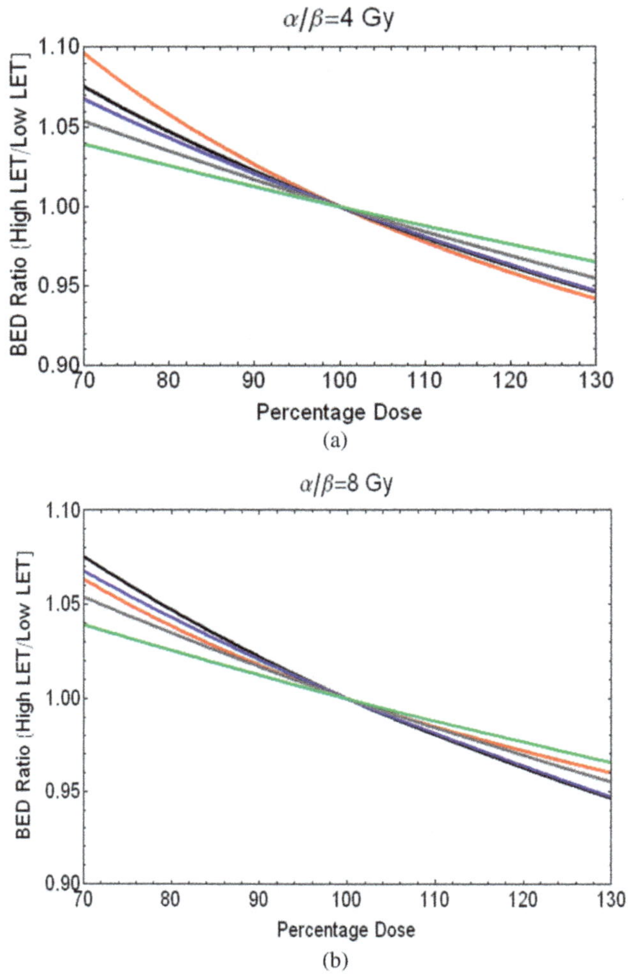

Figure 6.3. Ratios of high LET BED to low LET BED are plotted against percentage dose deviation (the factor x is expressed as a percentage), for a prescribed dose of 100%, and for different fraction numbers that provide iso-effective tumour control at the prescribed dose level for each fractionation schedule. The colour codes are red = 1#, black = 4#, blue = 10#, grey = 20# and green = 37#. The assumed RBE limits were RBE_{max} = 3, RBE_{min} = 1.6, and a low LET α/β ratio of 4 Gy (a) and 8 Gy (b), as shown appropriate for conditions in the end of a spread-out Bragg peak for protons in the context of (a) a slow growing tumour such as an adencocarcinoma and (b) a more rapidly growing human squamous cell cancer.

rapidly growing clonogenic cells, as may be found after repeated surgery and long intervals of time between the original diagnosis and definitive therapy. In contrast, the Swiss PSI centre has used protons with low dose per fraction, but in their case their development of beam scanning techniques permitted a more conformal physical dose distribution. All these results should eventually be subjected to rigorous comparative analysis using the BED with flexible RBE concept.

Major decisions are required in Europe before carbon ion dose per fraction is changed from a relatively fixed dose equivalent of around 3 Gy of photons.

The use of hypofractionation is increasing with newer photon treatments based on robotic guided linear accelerator techniques or multiple cobalt sources. The possibility that normal tissue late toxicities would be reduced by using hypofractionated PCP therapy compared to these other photon based techniques would be a reasonable starting point for clinical trials at some future time, but they would need to be guided by better and more specific data on the RBE, leading to improved BED estimates.

Present proton therapy practice requires some discussion here: the persistent use of a fixed proton RBE of 1.1, based on *in vivo* and *in vitro* experiments using rapid assays (Paganetti *et al* 2002, 2003), with a tendency to high α/β values, has a high chance of underestimating RBE values in late reacting tissues and in slowly growing tumours, as well as overestimating the RBE in some very rapidly dividing systems such as cancers of childhood; the latter could lead to under-dosage in long fractionated schedules (Jones *et al* 2012).

Iso-effective dose plots for protons are provided in chapter 9.

6.5 Effects of regions of higher and lower dose per fraction relative to the prescribed dose for different fractionation patterns

The medical prescription refers to a volume of interest, normally the planning target volume (PTV) where the prescribed dose often refers (e.g. in photon therapy) to the dose at the field intersection or other representative point such that the PTV is allowed to have a variation of dose by as much as −5% to +7%. Normally these percentages (expressed as 95% to 107% of the prescribed dose) do not deviate by these amounts, but with large field sizes and with difficult individual treatment geometry these figures are approached and rarely exceeded.

Homogenisation of dose has become increasingly possible due to intensity modulation in photon therapy and with similar approaches in PCP therapy due to beam scanning and weighting approaches. A deviation in dose will produce a changed biological effect, such that the surviving fraction will be multiplied by each fraction; in the case of the BED, which is derived as a logarithm, this allows the BEDs to be added for each fraction. For larger doses per fraction the biological deviation will be more significant, although fewer fractions are then given; for smaller doses per fraction the biological deviation will be smaller but will occur over a larger number of fractions.

In particular, if x represents the change in dose, then the low and high LET BED will be, respectively,

$$\mathrm{BED} = xD_{\mathrm{L}}\left(1 + \frac{xd_{\mathrm{L}}}{\left(\frac{\alpha}{\beta}\right)_{\mathrm{L}}}\right) \tag{6.14}$$

$$\mathrm{BED} = xD_{\mathrm{H}}\left(\mathrm{RBE}_{\max} + \frac{\mathrm{RBE}^2_{\min}\, x d_{\mathrm{H}}}{\left(\dfrac{\alpha}{\beta}\right)_{\mathrm{L}}}\right) \tag{6.15}$$

where, for example, $x = 0.96$ for 96% of the prescribed dose (of 100%), and 1.05 for 105% of prescribed dose, etc.

It is instructive to compare BED values for various values of x in low and high LET radiations, for two tumours or tissues with $\alpha/\beta = 4$ Gy and $\alpha/\beta = 8$ Gy in figures 6.3(a) and (b). A nonlinear effect is apparent, probably caused by the complexity of PCP radiobiology, where the RBE is inversely related to dose, with RBE 'crossing-over' effects.

These high LET dose changes tend to oppose under-dose (since the BED ratios exceed unity at doses below the prescribed dose) and also protect against over-dose (where the ratios are lower than unity), although only partial protection is given.

So, high LET can be considered to be more 'forgiving' in these respects. However, when figures 6.3(a) and (b) are looked at closely, it can be seen that the α/β values change the order of these BED ratio plots for different fraction numbers, indicating the need for individual calculations of the BEDs in all cases.

It is important to note that the dose gradient across the tumour target volume may be different between an x-ray (photon) plan and a PCP plan, where greater uniformity can be achieved, although this is based on assumptions linking LET and RBE, as in the work of Kanai *et al* (1997).

In Japan, where, from the example of a single field, the distal dose is down-weighted because of the increased LET due to a greater number of Bragg peaks whereas the proximal part of the target contains more plateau dose regions with lower LET. Ideally, comparison of an averaged BED across the target volume should be used to compare high and low LET target volumes. Failure to do so may allow an advantage for either modality. For example, schematic proton (red line) and x-ray (green curve) dose distributions across a target volume are shown in figure 6.4. There is a dilemma as to what proton dose should be given to provide the same tumour control; should the proton dose be matched to the calculated average x-ray dose across the volume (as attempted here), rather than be close to the minimum or maximum x-ray dose. Techniques such the uniform equivalent dose of Niemierko (1997) may be of use in this dilemma.

6.6 Taking RBE uncertainty into account in fractionation

A suitable method, using the index S_{I}, was given in chapter 3. A further example is given here.

By applying 25% errors to equation (3.8) in chapter 3, then

$$S_{\mathrm{I}} = \frac{P_{\mathrm{H}}}{P_{\mathrm{L}}}\left(\frac{\mathrm{RBE}_{\mathrm{norm}}}{\mathrm{RBE}_{\mathrm{tum}}}\frac{1+0.25}{1-0.25}\right) = \frac{P_{\mathrm{H}}}{P_{\mathrm{L}}}\left(\frac{\mathrm{RBE}_{\mathrm{norm}}}{\mathrm{RBE}_{\mathrm{tum}}} \times 1.67\right) \tag{6.16}$$

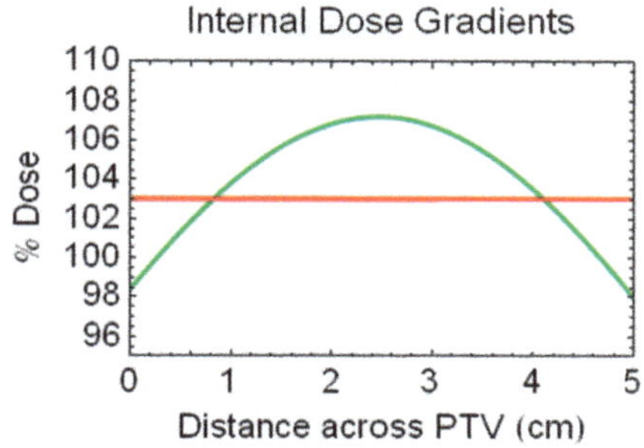

Figure 6.4. Percentage dose across a PTV of width 5 cm for protons (red) and x-rays (green). This refers to a hypothetical treatment where there is non-uniformity of dose in the x-ray case and excellent uniformity of dose for the proton beam, although non-uniformity of LET may influence the proton effectiveness, as given in by the ratios presented in figure 6.3. Reproduced from Jones *et al* (2012).

Accordingly, it is necessary to improve P_H (the physical sparing ratio of the high LET radiation) by 1/1.67 = 0.6, approximately, in order to maintain the same value of S_I, which would be a further challenge to the treatment planning physicist!

The situation is more difficult than the above for important normal tissues (or organ at risk, OAR) that extend into the PTV (in this case $P_H = P_L$). For example, consider the optic chiasm (which is the connection between eyes and brain) within the PTV of an adjacent tumour of the skull base, pituitary or supra-sellar region. The dose received by the OAR is then assumed to be the same as the tumour in the case of low LET radiation. but if the OAR has a higher RBE than the tumour, then the OAR will be susceptible to radiation injury, depending on its tolerance level. The only way of reversing this adverse RBE condition would be to lower the BED by changing the dose fractionation. In conventional low LET x-ray (photon) radiotherapy, the standard practice is to use a large number of fractions (often 30 or more). However, hyperfractionation of a high LET treatment will lead to high RBEs; the alternative course of action would be to consider the opposite—hypofractionation—since the high LET RBE is then lowered. Indeed, the RBEs between the two systems (tumour and OAR may reverse, as shown previously in figure 6.3.

However, if the dose can be more selectively deposited in the necessary target volumes without clinically significant alteration in surrounding tissue function, then optimisation systems show that a higher dose per fraction can be of benefit, it is economically more desirable and is easier for the patient (Jones and Dale 1998).

Worked example of fractionation effects for critical normal tissue included in PTV

A normal tissue such as optic chiasm with $\alpha/\beta = 2$ Gy falls within the PTV of a slow growing tumour with $\alpha/\beta = 4$ Gy. Assume tumour the RBE limits are 4 and 1.4, with normal tissue RBE limits of 6 and 1.25, respectively. Ideally the tumour should receive a dose equivalent to 72 Gy in 36 fractions of megavoltage x-rays and the tolerance of the chiasm is 50 Gy in 25 fractions with $\alpha/\beta = 2$ Gy.

Tolerance BED = (50 (1 + 2/2) = 100 $Gy_{[2]}$.

Intended tumour BED = 72(1 + 2/4) = 108 $Gy_{[4]}$.

The estimated doses per fraction to preserve a tumour iso-effect equivalent to 72 Gy in 36 fractions for 5, 15, 25 and 35 fractions are 3.71, 1.52, 0.97 and 0.71 Gy, respectively. These schedules provide an OAR BED of 165.20, 163.60, 163.08 and 162.82 $Gy_{[2]}$, respectively.

It can be seen that the schedules with fewer fractions (and larger fraction sizes) produce the largest BED values in this case, so that the effect of a fall in RBE with dose per fraction does not 'rescue' the problem. Also, all BED values exceed tolerance. Also, if a lower dose then has to be given (at least to a part of the tumour adjacent to the OAR), it is necessary to check tumour BED values.

Then, for 1, 2, 3, 4, 5 and 6 fractions (considering the case of extreme hypofractionation) we have, to maintain an iso-effect for the normal tissue tolerance level, d_H doses of 8.11, 5.03, 3.74, 3.00, 2.51 and 2.17 Gy, respectively. These schedules give tumour BED values of 64.64, 65.10, 65.37, 65.56, 65.69 and 65.80 $Gy_{[4]}$, respectively. These are considerably less than the intended value of 102 $Gy_{[4]}$. It can be seen that, in this example, increasing the number of fractions improves the tumour BED. This is because the tumour α/β is larger than the OAR α/β.

If it is decided to undertreat part of the tumour PTV but accept a higher dose to the chiasm equivalent to 56 Gy in 28 fractions, then in 5, 10, 15, 20, 25, 30 and 35 fractions, to preserve this OAR iso-effect, the tumour control BED will be 73.50, 73.92, 74.12, 74.23, 74.30, 74.36 and 74.40 $Gy_{[4]}$. Again, it can be seen that increasing the fractionation is slightly better for tumour control, but it is interesting to note that the equivalent dose for conventional x-ray tumour control to 74 $Gy_{[4]}$ is 49.33 Gy in 2 Gy fractions.

Alternatively if the number of fractions is further increased to 40, and we accept a reduced tumour BED of say 85 $Gy_{[4]}$, this can be achieved by a dose per fraction of 0.5 Gy, but this gives 127.8 $Gy_{[2]}$ as a late effect BED. Another attempt might be to use 40 fractions to give the expected tolerance level of 100 $Gy_{[2]}$, which requires a dose per fraction of around 0.4 Gy, but then the tumour BED falls to be equivalent to 44.75 Gy in 2 Gy fractions.

So, the therapeutic window is not good and it would be reasonable not to proceed with particle therapy in this instance. In practice, a 'small amount' of tumour is sometimes given a reduced BED, sufficient to respect the tolerance of the chiasm.

6.7 The use of the LQ model with large fraction sizes

It is known that deviations from the LQ model occur in some cellular systems at doses above 6–8 Gy. Some modellers apply linear models above a certain threshold dose and extrapolate the effect to higher doses. There is no entirely satisfactory method to overcome this. However, it is important to realise that the increasing cell survival slope with dose provided by the LQ model can be considered to be an advantage when assessing normal tissue effects since the LQ inevitably provides the 'worst-case scenario' and is consequently protective on normal tissues. Deviations from the model matter less for the tumour with high α/β and for very high LET radiations since the cell survival curves have reduced curvature compared with low LET radiations.

6.8 Optimisation of fractionation using calculus methods

It is, in principle, possible to study tumour cell kill for an equivalent normal tissue toxicity grading or intensity level. A variety of mathematical techniques could be used such as numerical analysis or, most conveniently, graphical methods. However, relatively simple differential calculus (Jones and Dale 2000) can be used provided the discontinuous fraction number parameter n is replaced by a function of overall time (t) and the mean inter-fraction interval (f). The approximation

$$t = f(n - 1) \tag{6.17}$$

can be used to replace t and then n can be replaced by the normal tissue BED(NT) where

$$\mathrm{BED} = nd_{\mathrm{H}}\left(\mathrm{RBE}_{\max} + \frac{\mathrm{RBE}_{\min}^{2}\,d_{\mathrm{H}}}{k}\right) \tag{6.18}$$

where, for convenience, k is now the late reacting normal tissue α/β ratio, so that

$$n = \frac{\mathrm{BED}}{d_{\mathrm{H}}\left(\mathrm{RBE}_{\max} + \dfrac{\mathrm{RBE}_{\min}^{2}\,d_{\mathrm{H}}}{\alpha/\beta}\right)}. \tag{6.19}$$

The dose per fraction d_{H} will be used as d from now on.

With a repopulation correction factor included we know that

$$\mathrm{BED} = nd\left(\mathrm{RBE}_{\max} + \frac{\mathrm{RBE}_{\min}^{2}\,d}{\left(\dfrac{\alpha}{\beta}\right)_{\mathrm{tum}}}\right) - Kt. \tag{6.20}$$

Then, using the replacement given in equation (8.1), we obtain

$$\mathrm{BED} = nd\left(\mathrm{RBE}_{\max} + \frac{\mathrm{RBE}_{\min}^{2}\,d}{\left(\dfrac{\alpha}{\beta}\right)_{\mathrm{tum}}}\right) - Kf(n - 1) \tag{6.21}$$

and then replacing n by equation (8.3), and defining z as the tumour dose and d the normal tissue dose (where $d = gz$), leads to:

$$\mathrm{BED} = \frac{\mathrm{BED(NT)}}{(d(R_{\mathrm{max,NT}} + R_{\mathrm{min,NT}}d/k))}z\left(\mathrm{RBE}_{\mathrm{max}} + \frac{\mathrm{RBE}_{\mathrm{min}}^{2}z}{\left(\frac{\alpha}{\beta}\right)_{\mathrm{tum}}}\right) - Kf\left(\frac{\mathrm{BED(NT)}}{\left(d\left(R_{\mathrm{max,NT}} + \frac{R_{\mathrm{min,NT}}d}{k}\right)\right)} - 1\right). \tag{6.22}$$

Of course, if the normal tissues receive very low LET radiation, when the RBE factors will be close to unity, we can simplify as

$$\mathrm{BED} = \left(\frac{\mathrm{BED(NT)}}{(d(1 + d/k))} - 1\right)z\left(\mathrm{RBE}_{\mathrm{max}} + \frac{\mathrm{RBE}_{\mathrm{min}}^{2}z}{\left(\frac{\alpha}{\beta}\right)_{\mathrm{tum}}}\right) - Kf\left(\frac{\mathrm{BED(NT)}}{\left(d\left(1 + \frac{d}{k}\right)\right)} - 1\right). \tag{6.23}$$

The normal tissue dose per fraction symbol, d, can in all cases be replaced by gz. Then, a maximum tumour BED (which is proportional to tumour cell kill) that preserves the same normal tissue response occurs only when

$$\frac{\mathrm{d(BED)}}{\mathrm{d}z} = 0. \tag{6.24}$$

This last equation can be solved easily using mathematical software and is, very interestingly, independent of the actual normal tissue BED constraint, but does include the normal tissue α/β (given the symbol k). Plots of BED in this form show that optimum turnover points are found especially when the normal tissues sparing increases (low g values), as shown in figure 6.5.

Plots of the achievable tumour BED, which is iso-effective for the critical normal tissue, are shown as the estimated optimum z are shown in figure 6.5(a), where clear turnover points are seen as low doses. If the geometrical sparing factor (g) is further improved, as in figure 6.5(b), the tumour BEDs increase further and in this case the tumour repopulation rate is lower. The flatness of the BED plateau in some cases show that there is no optimum dose per fraction (LET > 1 when $g = 0.4$) in the example given. Generally, the BEDs increase with improvement of treatment geometry (a lower g value), and also with low α/β ratios, although higher repopulation rates will also reduce the BEDs.

For this set of parameters and the other assumptions made, hypofractionation (to very high doses) is predicted to be a better option for conditions of good normal tissue sparing and for more rapidly growing tumours. Such calculations can only be a rough guide and it would be essential to take further measures to ensure excellent

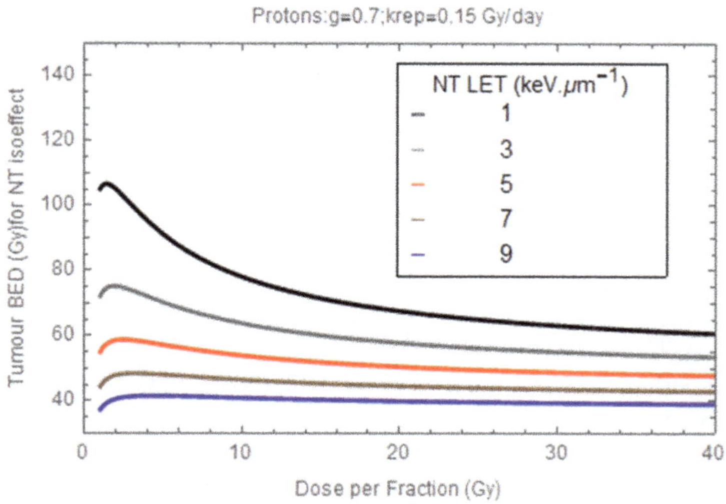

(a)

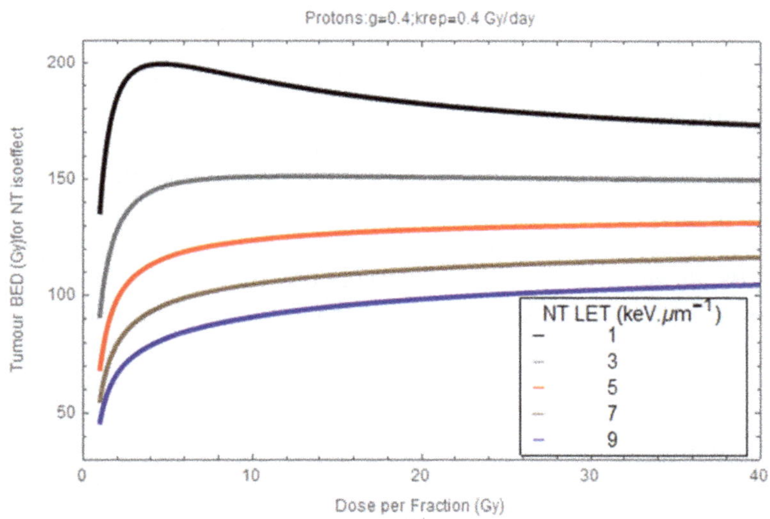

(b)

Figure 6.5. Plot of the tumour BED for an equal late normal tissue iso-effect against tumour dose per fraction using the equations given above and in chapters 2 and 9. In (a), for a geometric sparing factor (g) of 0.7, the optimum dose per fraction is around 2 Gy per day, but in (b), where $g = 0.4$, much higher tumour BEDs are achievable, but there is only an optimum dose per fraction for one condition, when LET = 1 keV μm^{-1}. For higher values of LET there is no optimum dose per fraction. Both cases assumed a tumour $\alpha/\beta = 10$ Gy.

results; the aim should always be give the *lowest* possible normal tissue dose consistent with the highest tumour BED. So, individual calculations will always be required. However, useful trends emerge such as the requirement of a higher dose per fraction for more rapidly growing tumours (with higher α/β ratios); the alternative would be to provide twice daily fractions using a lower dose per fraction (f is then shorter, around 0.78 days for say 10 fractions per week).

Further 3D plots with inclusion of the optimum tumour dose per fraction (iso-effective for the normal tissue of concern) will be of interest. The two examples given in figure 6.6 show, in figure 6.6(a), a broad front of optimal low dose per fraction (around 2 Gy), but which increases more sharply with increasing LET and with reducing *g* values, to high doses per fraction until there is no optimum value at lower values of *g*. The scope for an optimal dose per fraction is further reduced in figure 6.6(b) where the phase space occupied is far less due to the lower tumour α/β ratio (of 5 Gy) and repopulation equivalent (0.16 Gy d^{-1}).

A particular problem associated with the calculus method is that parameters used must represent the average values during treatment and cannot easily be changed during treatment. This creates a difficulty with accelerated repopulation, although this is only relevant perhaps to squamous cell cancers (SCCs) treated over durations longer than 28 days. Slower growing tumours have less significant repopulation factors and, in any case, ion beam therapy is normally completed well within this time frame. The present author considers it a mistake to assume that there is no repopulation up to 28 days, even in SCCs, for there is always cellular turnover even if this is at a lower rate.

6.9 Other contributions to fractionation

No account of fractionation would be complete without mention of these other important mechanisms that influence fractionation effects. Fractionation is often taught as being dependent on several factors beginning with 'R', including radio-sensitivity, repopulation, reoxygenation, repair and re-assortment (of the cell cycle). The α/β ratios used, if derived from clinical data, probably inherently include all these processes. We have already discussed repair and that α/β is the ratio of radio-sensitivities, along with the different susceptibility of α and β to changes in LET.

Reoxygenation rates are thought to be rapid in most tumours, with perhaps sarcomas and high-grade gliomas being exceptions. Much will depend on tumour and vascular supply responsiveness, but there is insufficient knowledge of the rate of reoxygenation in individual tumours at the present time. This key rate can in theory influence the potential outcomes of therapy, even in the case of positively charged particles (Dale and Jones 2007 (see book reference)). A single fraction offers no time for reoxygenation and even two fractions (given daily) will suffer from the fact that the second fraction will contain a greater fraction of hypoxic cells. The minimum time for significant although incomplete reoxygenation might be a week, consistent with four fractions given over 4–5 days. A greater number of fractions also might carry an advantage in the case of hypoxia caused by cyclical opening and closing of blood vessels in a tumour, since the volume of tumour treated during a hypoxic phase is expected to be less.

Re-assortment of the cell cycle was thought to be important, since high LET radiations exhibit less variation in radiosensitivity in the different phases of the cell cycle and so increasing fractionation would not be so important. This was established for neutron therapy but it is an effect that is mostly seen at low dose; higher doses that produce much lower surviving fractions appear to show less change with cell cycle position (see Catterall and Bewley 1979), presumably since the

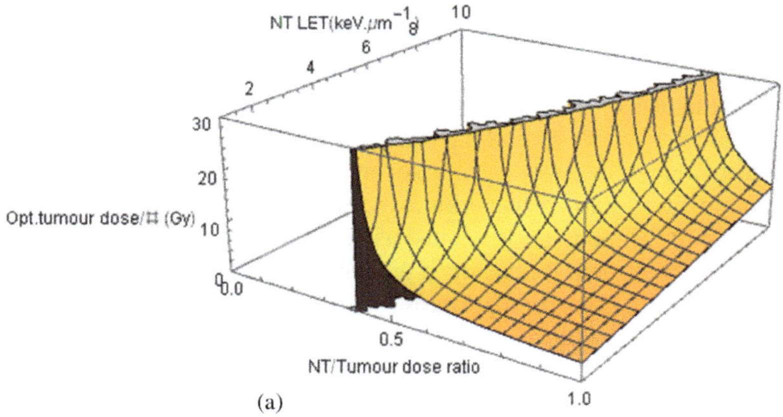

(a)

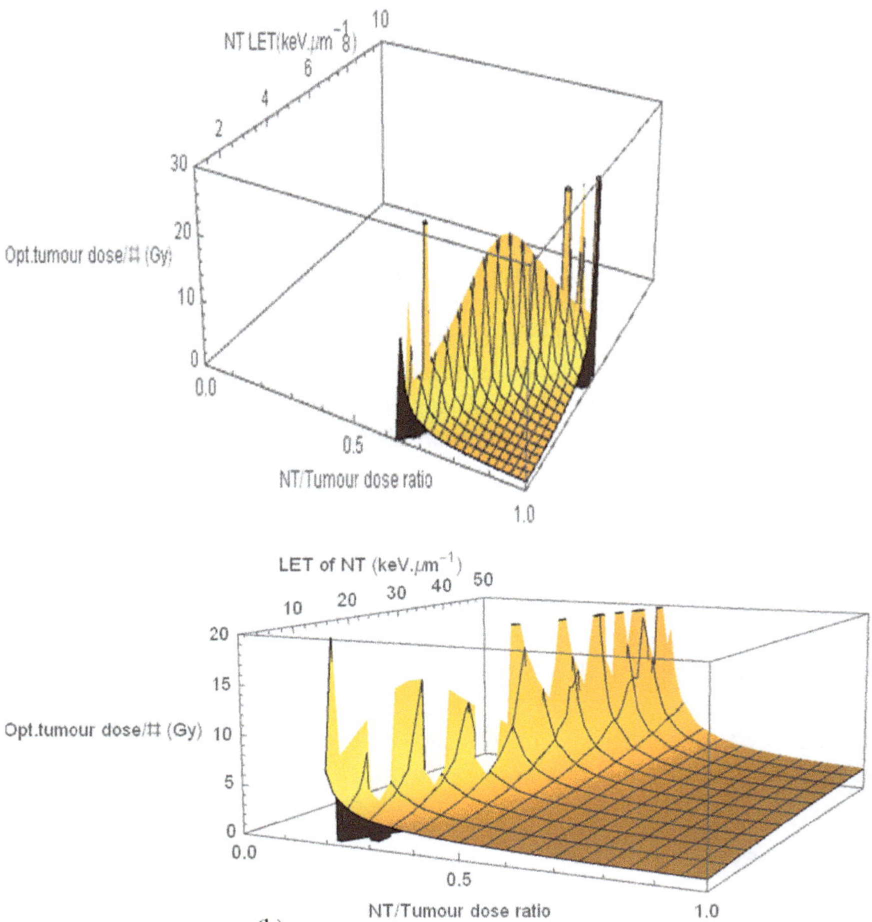

(b)

Figure 6.6. Plots of optimum tumour dose per fraction for maximum cell kill, against tumour low LET α/β ratio, for variations in normal tissue/tumour dose ratio, the sparing factor (g), for a constant BED of 100 $Gy_{[2]}$ in the normal tissue of interest, assuming five fractions per week are given. The tumour α/β ratios and repopulation equivalents were assumed to be (a) 10 Gy and 0.6 Gy d^{-1} and in (b) 5 Gy and 0.15 Gy d^{-1}, respectively.

cell kill then follows the type B (β-related) cell kill process, which has a lower RBE (the RBE_{min}). Such an effect will apply to both normal tissues and tumours.

The results obtained by Fowler and colleagues, in terms of an improvement in tumour control for x-rays by extending the number of fractions beyond a single fraction, were consistent with cell cycle re-assortment and reoxygenation, in a tumour where reoxygenation was known to be almost complete in three days after a large single dose. The use of fast neutrons gave just as high tumour control as the optimum dose per fraction and treatment time, even if given in a single fraction because of their reduced dependency on hypoxia and perhaps to a lesser extent on cell cycle position.

Repopulation factors can be used for unintended treatment gap corrections, as discussed previously, and included in overall BED calculations for longer treatment schedules. However, since most ion beam treatments do not at present extend beyond a month, this is considered unnecessary in most tumour types. However, the use of ion beams as a boost after IMRT will usually cause treatment time to be prolonged.

What if radiobiological parameters change during irradiation? We assume that the parameters used in fractionation calculation are the average values during radiotherapy. Where abrupt changes occur it may be necessary to use different forms of mathematics such as series expansions.

The effect of irradiated normal tissue volumes has only been considered briefly above; the precise location of the radiation relative to the 3D vascular supply of organs may yield further useful data that will permit more selective hadrontherapy with reduction in normal tissue complications.

Another 'contribution' to the effective BED values would be those of chemotherapy, age, surgery, etc, which can provide a further increment in cell kill or susceptibility to radiation effects. In principle, these can be represented by a variable BED increment, as shown elsewhere (Jones *et al* 2006).

One further important issue is *malignant induction*. It has been argued that the most protective effect of PCP therapy is a reduction in normal tissue cell numbers exposed to radiation, especially since RBE effects can increase the risk of malignant induction in individual cells. This effectively means that hadrontherapy should only use the lowest number of treatment fields, which meets with the treatment aims and tissue constraints. In principle, fractionation may alter the yield of malignant cells in some hypofractionated cases if the 'turnover points' of malignant induction rates are exceeded and cell killing rather than malignant induction then dominates. Much further work is required in this area, although the changes are likely to be small and must be assessed together with any detriments in tumour control or normal tissue side effects that could occur (Jones 2009, Timlin *et al* 2011, Timlin *et al* 2015).

6.10 Summary

Fractionation of high LET PCP therapy imposes challenges and opportunities to improve and optimise the therapeutic index and cost effectiveness, given that hypofractionation, even if applicable in only some clinical situations, will allow greater patient throughput. This can be achieved by further experimental RBE and fractionation programmes, eventually leading to carefully constructed and conducted clinical trials that will compare PCP fairly with low LET radiotherapy alternatives.

References

Barendsen G W and Broerse J J 1970 Exprimental radiotherapy of a rat rhabdomyosarcoma with 15 meV neutrons and 300 keV x-rays. effects of fractionated treatments applied five times a week for several weeks *Eur. J. Cancer* **6** 89–109

Carabe-Fernandez A, Dale R G and Jones B 2007 The incorporation of the concept of minimum RBE (RBE_{min}) into the linear-quadratic model and the potential for improved radio-biological analysis of high-LET treatments *Int. J. Radiat. Biol.* **83** 27–39

Catterall M and Bewley D K 1979 *Fast Neutrons and The Treatment of Cancer* (London: Academic) pp 92–4

Dale R G 1989 Time-dependent tumour repopulation factors in the linear quadratic equations – implications for treatment strategies *Radiother. Oncol.* **15** 371–82

Douglas B G and Fowler J F 1976 The effect of multiple small doses of x-rays on skin reactions in the mouse and a basic interpretation *Radiat. Res.* **66** 401–26

Denekamp J 1973 Changes in the rate of repopulation during multifraction irradiation of mouse skin *Brit. J. Radiol.* **46** 381–7

Fowler J F, Sheldon P W, Denekamp J and Field S B 1976 Optimum fractionation of the C3H mouse mammary carcinoma using x-rays, the hypoxic-cell radiosensitizer Ro-07-0582, or fast neutrons *Int. J. Radiat. Oncol. Biol. Phys.* **1** 579–92

Fowler J F 1981 *Nuclear Particles in Cancer Treatment* Medical Physics Handbook (Bristol: Adam Hilger)

Fowler J F 1984 The first James Kirk memorial lecture. What next in fractionated radiotherapy? *Brit. J. Cancer Suppl.* **6** 285–300

Fowler J F 1989 The linear quadratic formula and progress in fractionated radiotherapy *Brit. J. Radiol.* **62** 679–94

Fowler J F 2010 21 years of biologically effective dose *Brit. J. Radiol.* **83** 554–68

Fowler J F, Denekamp J, Sheldon P W, Smith A M, Begg A C, Harris S R and Page A L 1974 Optimum fractionation in x-ray treatment of C3H mouse mammary tumours *Brit. J. Radiol.* **47** 781–9

Grassburger C, Trofimov A, Lomax A and Paganetti H 2011 Variations on linear energy transfer within clinical proton therapy fields and the potential for biological treatment planning *Int. J. Radiat. Oncol. Biol. Phys.* **80** 1559–66

Joiner M C 2004 A simple alpha/beta-independent method to derive fully isoeffective schedules following changes in dose per fraction *Int. J. Radiat. Oncol. Biol. Phys.* **58** 871–5

Jones B 2009 Modelling carcinogenesis after radiotherapy using Poisson statistics: implications for IMRT, protons and ions *J. Radiol. Prot.* **29** A143–57

Jones B, Carabe-Fernandez A and Dale R G 2006 Calculation of high-LET radiotherapy dose required for compensation of overall treatment time extensions *Brit. J. Radiol.* **79** 254–7

Jones B, Carabe-Fernandez A and Dale R G 2007 The oxygen effect *Radiobiological Modelling in Radiation Oncology* (London: British Institute of Radiology) ch 8

Jones B and Dale R G 1998 Radiobiologically based assessments of the net costs of fractionated focal radiotherapy *Int. J. Radiat. Oncol. Bio. Phys.* **41** 1139–48

Jones B and Dale R G 2000 Estimation of optimum dose per fraction for high LET radiations: implications for proton radiotherapy *Int. J. Radiat. Oncol. Biol. Phys.* **48** 1549–57

Jones B and Dale R G 2008 Radiobiological compensation of treatment errors in radiotherapy *Brit. J. Radiol.* **81** 323–6

Jones B, Dale R G, Deehan C, Hopkins K I and Morgan D A L 2001 The role of biologically effective dose (BED) in clinical oncology *Clin. Oncol.* **13** 71–81

Jones B, Dale R G and Gaya A 2006 Linear quadratic modelling of increased late normal tissue effects in special clinical situations *Int. J. Radiat. Oncol. Biol. Phys.* **64** 948–53

Jones B, Dale R G and Kakhsar S J 2000 Biological equivalent dose assessment of the consequences of hypofractionated radiotherapy *Int. J. Radiat. Oncol. Biol. Phys.* **47** 1379–84

Jones B, Underwood T C, Carabe-Fernandez A and Dale R G 2011 Further analysis of fast neutron relative biological effects and implications for charged particle therapy *Brit. J. Radiol.* **84** 11–8

Jones B, Wilson P, Nagano A, Fenwick J and Mckenna G 2012 Dilemmas concerning dose distribution and the influence of relative biological effect (RBE) in proton beam therapy of medulloblastoma *Brit. J. Radiol.* **85** 912–8

Haviland J S *et al* 2013 START Trialists' Group The UK Standardisation of Breast Radiotherapy (START) trials of radiotherapy hypofractionation for treatment of early breast cancer: 10-year follow-up results of two randomised controlled trials *Lancet Oncol.* **14** 1086–94

Hopewell J W, Millar W T and Lindquist C 2012 Radiobiological principles: their application to γ knife therapy *Prog. Neurol. Surg.* **25** 39–54

Kanai T, Furusawa Y, Fukutsu K, Itsukaichi H, Eguchi-Kasai K and Ohara H 1997 Irradiation of mixed beam and design of spread-out Bragg peak for heavy-ion radiotherapy *Radiat. Res.* **147** 78–85

Karube M, Yamamoto N, Nakajima M, Yamashita H, Nakagawa K, Miyamoto T, Tsuji H, Fujisawa T and Kamada T 2016 single-fraction carbon-ion radiation therapy for patients 80 years of age and older with stage I non-small cell lung cancer *Int. J. Radiat. Oncol. Biol. Phys.* **95** 542–8

Lea D E 1962 *Action of Radiation on Living Cells* 1st edn 1946, 2nd edn 962 (Cambridge: Cambridge University Press)

Niemierko A 1997 Reporting and analyzing dose distributions: a concept of equivalent uniform dose *Med. Phys.* **24** 103–10

Paganetti H 2014 Relative biological effectiveness (RBE) values for proton beam therapy. Variations as a function of biological endpoint, dose, and linear energy transfer *Phys. Med. Biol.* **59** R419–72

Paganetti H, Niemierko A, Ancukiewicz M, Gerweck L E, Goitein M, Loeffler J S and Suit H D 2002 Relative biological effectiveness (RBE) values for proton beam therapy *Int. J. Radiat. Oncol. Biol. Phys.* **53** 407–21

Pop L A, Millar W T, van der Plas M and van der Kogel A J 2000 Radiation tolerance of rat spinal cord to pulsed dose rate (PDR-) brachytherapy: the impact of differences in temporal dose distribution *Radiother. Oncol.* **55** 301–15

Timlin C, Houston M and Jones B 2011 Malignant induction probability maps for radiotherapy using x-ray and proton beams *Brit. J. Radiol.* **84** S70–8

Timlin C, Warren D R, Rowland B, Madkhali A, Loken A, Partridge M, Jones B, Kruse J and Miller R 2015 3D calculation of radiation-induced second cancer risk including dose and tissue response heterogeneities *Med. Phys.* **42** 866–76

Thames H D and Hendry J H 1987 *Fractionation in Radiotherapy* (London: Taylor and Francis)

Chapter 7

The case for using a variable proton RBE

7.1 Introduction

In standard practice, the delivered proton tumour dose is the intended photon equivalent physical dose divided by the relative biological effect (RBE), so if RBE allocations to different tissues and tumours are incorrect, there will be inevitable under- or over-dosage in the tumour and/or in the normal tissues. The clinical significance of this will vary due to a variety of reasons such as the physical dose, treatment volume, local anatomy, medical history and other factors that may change normal tissue tolerances or influence tumour control. The degree of dose variation caused by an incorrect RBE assignment can not only exceed those in the ICRU prescription guidelines (of −5% to +7% of the prescribed dose), but also those allowed by legal requirements for dose precision in most countries of around 2% (Dale *et al* 2009, ICRU 2010). This is a sensitive issue in the clinic, since there are implications that tissue doses may be inappropriately high if the proton RBE is assumed to be 1.1, but the actual tissue RBE exceeds this number. This is the presently accepted international standard of practice. This policy of using a constant RBE of 1.1 in all tissues, tumours, and at all levels of dose and linear energy transfer (LET), is endorsed in ICRU 78, but has been questioned by radiobiologists for some time, with recent calls for its reassessment, for example by Blomquist *et al* (1993), Jones and Dale (2000), Paganetti (2014), Tommasino and Durante (2015), Dasu and Toma-Dasu (2013), Jones (2015 and 2017) and Jones (2016). This chapter is based on this last reference (where further details can be obtained) and provides the rationale for rejecting the present standard policy. Alternative approaches are presented in chapters 8 and 9.

The decision to adopt this single value was made on the basis of some well-intended radiobiological experiments. It also followed the precedent of using a constant RBE in UK fast neutron therapy, despite extensive evidence that RBE varied with dose per fraction (Hornsey and Field 1974), although this was in the era before the linear-quadratic (LQ) model was used extensively. Subsequent studies of these data sets in a

doi:10.1088/978-0-7503-1338-4ch7

LQ setting show that the extent of the variation in RBE with dose per fraction depends on the control low LET α/β ratio in a dose-dependent manner (Carabe-Fernandez *et al* 2007, 2010, Jones *et al* 2011a), with examples shown in chapter 5. When these decisions were taken, the treatment planning was not computerised and so could not have accommodated a variable RBE easily, and the easier alternative option of reallocating tissue tolerances based on higher RBE values was not adopted, which may in part have been due to the lack of 3D dose distributions.

This chapter considers the reasons why the proton RBE must be a variable rather than a constant, by referring to basic radiobiological principles within the overall framework of the LQ model. Increased clustering of ionisation events with increasing LET produces an inevitable higher proportion of unrepairable damage, conferring increased radiosensitivity and reduced fraction sensitivity compared with lower LET radiations. It has been misleading to define a boundary between 'low' and 'high' LET since protons that produce increasing RBE occur at 'low' LET values (from 1–2 keV μm^{-1} as encountered in most spread-out Bragg-peaks (SOBPs)), and more so at values of 2–10 keV μm^{-1} (Belli *et al* 2000, Britten *et al* 2013). In the present chapter, high LET will refer to any LET that exceeds the control megavoltage photon or electron low LET value of around 0.22 keV μm^{-1}. The literature is littered with experiments using kilovoltage x-rays with LET values of 1 keV μm^{-1} or more; these result in lower RBE values (often less than 1.0, which implies that the control radiation LET used is higher than the test irradiation), a situation that does not arise if megavoltage control irradiations are used (see also chapter 1). There is a steep increase in the RBE beyond this value, and more so in the case of protons (Belli *et al* 2000, Britten *et al* 2013) than other ions, where the LET and RBE gradient is less marked, as shown in chapters 8 and 9.

7.2 Arguments to preserve the status quo or avoid using RBE

Many authorities have claimed that there is no indication of a need to change the present system, since the clinical results are satisfactory, but there are few, if any, publications that can justify this, whereas examples of 'unexpected' toxicities have been a cause of concern to referring physicians. It is relevant to mention the experience in the UK where two radiotherapy centres under- and over-dosed patients for a number of years (due to 60-cobalt dose rate errors). In neither of these centres did the clinicians notice either lack of effectiveness or enhanced complications, which were found when formal retrospective analysis was made. It should be incumbent on proton centres, or indeed any centre using new technologies, to report their data annually. It is salutary to note that when the British government purchased a large stock of radium and established radium institutes in most major cities in the first half of the twentieth century, it was mandatory to report all results on an annual basis, a practice which persisted until the 1970s. This may seem excessive in present times, but it is in the public interest.

Others have argued that RBE values other than 1.1 should perhaps not be used, because the RBE is so 'complex', and that LET mapping may be sufficient, with some form of dose reduction.

Another suggestion has been to use the product of dose and LET; this may be useful in situations where dose and LET are both increased, but since the RBE falls as dose increases, this index will be nonlinear, and must be vary considerably in the provision of effect thresholds for different tissues and tumours. It is self-evident that when the dose is slightly reduced to be just below the threshold for toxicity, but LET is raised (slightly), toxicity could occur despite the product being unchanged. This follows due to the increase in the RBE with both LET and the fall of dose per fraction. In terms of physical dimensions, this product also does not seem to refer to any useful physical concept. For further information see chapter 4. Although such an approach may be helpful in the short term, it should be applied with caution. This is because should any dose modification factor be used based directly on LET, it will need to be adjusted for the dose itself, and the factor will inevitably vary with the tissue or tumour being considered. Also, changing the dose per fraction, such as use of hypofractionation, would require quite different parameters. The overall system will turn out to be nonlinear and a masquerade for the RBE. The RBE is a physical reality that cannot be ignored. The situation can be compared with, for example, using the constant acceleration due to gravity at the surface of the Earth with good results for falling bodies, but then applying this universally; on the Moon or Mars it will be quite different. Better by far would be to have a general system for establishing reasonable values of the RBE in all likely biological/medical situations, based on reasonable understanding of the underlying science.

7.3 Justification of a variable RBE

The justification for a variable rather than a constant proton RBE is considered from considerations of standard radiobiological principles, beginning with the definition of the RBE and how the ratio changes for different cell/tissue types, which with other physical reasons contribute to the multifactorial nature of the RBE, as presented in the *British Journal of Radiology* by Jones (2016). Then two proton data sets are analysed in further detail than in the original publications.

(a) A further analysis of the data of Britten *et al* (2013), to find the relationship between the RBE and dose per fraction at various depths in the beam. The published radiosensitivities of the control and test cells were used to provide RBE_{max} and RBE_{min} values, which were then used in equation (2.19) (chapter 2) to provide the RBE plots.

(b) A reanalysis of the data of Paganetti *et al* (2002). This data set influenced the decision to continue with the 1.1 RBE to the present time.

The above points are now considered.

7.3.1

The definition of the RBE refers to its measurement as the low LET (reference) dose divided by the high LET dose,

$$\text{RBE} = \frac{\text{dose of low LET radiation (as reference)}}{\text{dose of high LET radiation}}$$

when the same bio-effect is achieved by each quality of radiation (i.e. when both are iso-effective). The numerator dose is depends on repair proficiency to a greater extent than the denominator, due to the greater proportion of unrepairable damage with increasing LET. To meet iso-effective conditions for a specified denominator high LET dose, the required low LET numerator dose per fraction will be much larger in the case of a more radioresistant cell/tissue (with high repair capacity) than in the case of a more radiosensitive cell/tissue system (with lower repair capacity). In this way slower growing/more radioresistant systems (which have large fraction sensitivities associated with low α/β ratios) will have the highest RBEs. Conversely, cells that are highly radiosensitive, with a high α/β ratio, will require a lower numerator dose and the high LET dose will not differ by as much, since the absence of repair proficiency will not be so influential where there is a higher proportion of unrepairable damage. Consequently, the RBE will be lower and can approach 1.0 in highly radiosensitive repair-deficient mutant cells. For example, the data of Weyrather *et al* (1999) show that the low LET radiosensitivity correlates inversely with the RBE. The RBE ratio consequently depends mostly on the intrinsic radiosensitivity/repair proficiency of the cell/tissue type, being scaled mostly by the numerator dose. It follows that in the more radiosensitive/rapidly growing cells/tissues, or with reduced DNA repair capacity, the changes in the RBE with dose per fraction will be small. Since repair capacities and radiosensitivities are determined by a large number of genes, then the RBE must vary. This is similar to population frequency distributions of height, body weight, blood pressure, etc, where multifactorial influences cause variation.

7.3.2

Another reason is provided by the mathematical boundary conditions for the RBE. These extreme RBE limits are defined by radiosensitivity ratios of the higher LET radiosensitivity state divided by the control LET radiosensitivity. This is represented by the separate nonlinear increments in α and β that occur with increasing LET. Derivations are provided in the footnote[1]. Fast neutron experiments are appropriate for simulation of the proton RBE in the Bragg peak region, since most ionisation occurs due to the formation of recoil protons. Examples are provided in the fast neutron data (see, e.g. Carabe-Fernandez *et al* (2007, 2010), Warenius *et al* (1994), with further examples in chapter 5), and the proton data of Britten *et al* (2013), where α increases with LET to a much greater extent than β. The maximum RBE is

[1] Using suffixes L and H for low and a higher LET respectively, iso-effective doses can be represented as:

$$\alpha_L d_L + \beta_L d_L^2 = \alpha_H d_H + \beta_H d_H^2.$$

As d_H approaches zero, β terms can be neglected, so the RBE (the ratio d_L/d_H) becomes α_H/α_L, denoted as RBE_{max},

If d_L becomes large, α terms are neglected, so the RBE approaches $\sqrt{\beta H}/\sqrt{\beta L}$, denoted as RBE_{min}.

It follows that at all intermediate doses, RBE will have a value between those of RBE_{max} and RBE_{min}.

then given by α_H/α_L, found at very low dose since the α radiosensitivity parameter dominates cell killing at low doses. The minimum RBE is given by $\sqrt{\beta_H}/\sqrt{\beta_L}$ at very high dose where the β component of cell kill dominates. For all doses between these limits, the RBE must be of a variable intermediate value, which depends on dose itself (due to the changing contribution of α and β to cell kill with dose). Also, the intrinsic variability of the low LET α and β values, which increase nonlinearly with LET, will further contribute to RBE variations.

7.3.3

It is commonly assumed that because the measured proton RBE in experiments is so close to unity then further deviations are insignificant, especially if compared with ion beams having higher RBEs of around 3. This rationale is often used to support the use of a constant proton RBE on the grounds that small deviations either side of 1.1 might not be significant in clinical practice, especially since the megavoltage photon control radiation by definition has an RBE of 1. This simplistic argument can be counteracted by considering the effect of a change in dose on the biological effective dose (BED). For further details see Jones (2016), where it is shown that any error is just as significant regardless of the RBE value chosen.

7.3.4

The RBE will depend on the following physical factors:

- particle charge (z) and energy,
- depth/position within patient,
- treatment volume (due to superimposition of low and high LET regions off and within Bragg peaks, respectively).
- beam contamination with neutrons and γ radiation.

For all these reasons, it follows that the RBE will be a continuous variable. There will be an infinite number of RBEs in any practical treatment plan, just as with dose, although for convenience dose and RBEs can be binned into ranges.

7.3.5

Two experimental data sets.

7.3.5.1 The Britten et al *(2013) data set*

Most proton RBE experiments have measured the RBE using single fraction experiments conducted in the mid-SOBP region of a single field. Higher LET and RBE values will inevitably occur towards the end of SOBP range. The exemplary study of Britten *et al* (2013) performed radiosensitivity and RBE estimations at different depth/ranges and LET values. Changes in the α radiosensitivity parameter with LET in human Hep-2 cells are shown in figure 7.1(a), with values dependent on the initial incident energy. There are insufficient data points to determine the true position of the α-radiosensitivity and LET turnover points, but the increase in α appears to be inversely related to the incident energy. Only the lower incident energy results appear to be similar to the data of Belli *et al* (2000), where turnover occurs at

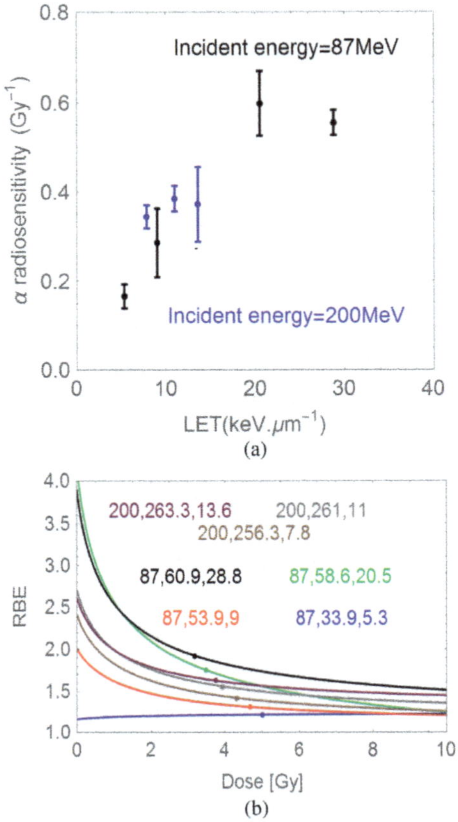

Figure 7.1. Further analysis of data of Britten *et al* (2013). (a) For α radiosensitivity in Hep-2 cells with changes in LET, for two different values of the initial incident energy. (b) RBE plots with dose per fraction calculated using equation (2.19) (chapter 2), constructed from the α and β radiosensitivities reported in the data of Britten *et al* (2013) for Hep-2 human cells, at various positions in SOPBs and obtained using two different incident energies. For each colour, the set of three numbers respectively refer to: incident energy (MeV), depth (mm) and LET (KeV μm^{-1}). The superimposed points are RBEs for a 0.1 surviving fraction. Reproduced from Jones (2016).

an LET value of around 30.5 KeV μm^{-1}. The RBE variation with depth and dose per fraction in this study is shown in figure 7.1(b) for each incident energy. These curves show the large variation of RBE with dose per fraction and LET in different parts of a proton beam. The points shown in figure 7.1(b) are the published RBE values for a surviving fraction of 0.1; these fit on each curve almost exactly, although the curves were not fitted by their direct use, thus increasing the overall confidence in the high LET LQ model when modified with RBE_{max} and RBE_{min} parameters. The Kolmogorov–Smirnov test statistic for goodness of fit of these points approaches 1, the prediction accuracy average being 0.998 ± 0.002.

7.3.5.2 The Paganetti et al (2002) data sets

Linear regression models (with standard error weighting where possible), have been used to provide a linear fit of the form RBE = $m \times$ dose + c, where m is the slope and c

the intercept. Also, the LQ model in its BED form was used to define iso-effective conditions for plots of RBE against dose per fraction, using least-squared nonlinear fitting techniques. Statistical comparisons were made using absolute residual values comparing observed and expected data results, the summated chi-squared statistic and the distribution free Kolmogorov–Smirnov test. The data sets were analysed separately for (a) all data, and (b) data where the RBE exceeds unity, which eliminates all experiments where low keV photon controls were used, since the latter may have a higher LET than the proton LET in the SOBP, thus causing an 'inverted' RBE ratio, with a value below one. Although this may introduce a bias, there is a good scientific *a priori* reason for not using such data; indeed the original data set includes a reverse bias by inclusion of these experiments. For preclinical experiments, the control irradiation should always be megavoltage photons or very well filtered 250 KeV x-ray beams (see chapter 1). Data with equivocal or no standard error estimates were also excluded from the present analysis, unlike in Paganetti *et al* (2002).

Both *in vitro* and *in vivo* data sets show a modest increase in mean and median values of RBE when RBE values below unity are excluded, as shown in table 7.1. For this reason the further analysis contains only the RBE results greater than unity, and excludes experiments where the control radiation used relatively low keV photons in the photoelectric energy range. The *in vitro* data contains four such exclusions and the *in vivo* data fourteen exclusions.

7.3.6

In vitro assays

In the published series there are considerable variations in proton RBE. Some data points at low dose per fraction substantially exceed 1.1, with small error bars. However, a greater number of experiments were performed at high dose per fraction where lower RBEs predominate. These experiments can be criticised from the radiotherapy standpoint since most of the cell lines tested are from the Chinese hamster ovary, with lower chromosome numbers than in somatic human cells. *In vitro* cell growth conditions also favour proliferation, which confers increasing photon radiosensitivity, which will drive down the RBE. The range of cells used is very limited and based on low cost assays, rather than using a pool of human cell lines with a wide spectrum of cell kinetics and intrinsic radiosensitivities. Even so,

Table 7.1. Comparisons of means and median values of data characterised as having RBE > 0 and RBE > 1.

	In vitro RBE	Dose (Gy)	*In vivo* RBE	Dose (Gy)
Data RBE > 0				
Mean (SD)	1.21 (0.20)	4.8 (2.69)	1.08 (0.14)	12.2 (9.02)
Median	1.20	5.37	1.05	12.4
Data RBE > 1				
Mean (SD)	1.22 (0.18)	5.0 (2.68)	1.12 (0.07)	8.7 (8.04)
Median	1.23	5.6	1.12	10.07

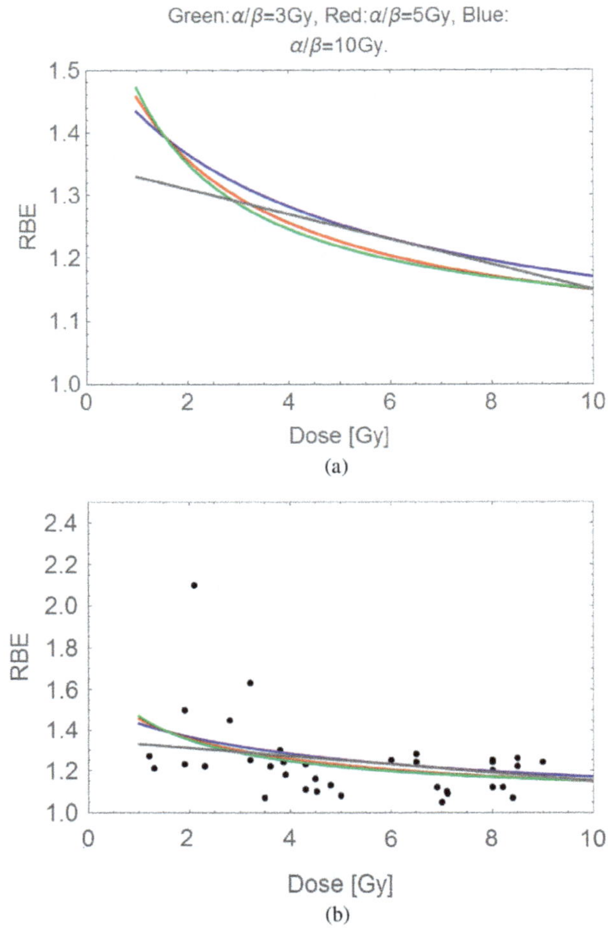

Figure 7.2. *In vitro* data (RBEs > 1 only) of RBE and dose data derived from Paganetti *et al* (2002), with modelled plots of dose per fraction and RBE plots using the weighted least-squares fitted 1.35–0.02d linear regression model (the straight line), and the LQ model, for varying α/β ratios (which provides curves): (a) magnified view of plotted models and (b) with superimposed data over a larger scale using the same colour codes. Due to frequent overlapping, error bars are omitted, but are in the original reference. The LQ RBE plots with dose per fraction are constructed using equation (6.5)[2]. Reproduced from Jones (2016).

the data does show an inverse relationship with dose, which may be fitted by simple-linear, or LQ model functions (figures 7.2(a) and (b)).

7.3.7
In vivo assays

In general, *in vivo* tissue assays can be divided into two types: acute and late tissue reactions. Each have different mechanisms. The acute reaction is due to cellular

[2] This leads to an RBE given by $1/2(-k + 4d_H\, k\, \text{RBE max} + 4d_H^2\, \text{RBE min}^2)/d_H$.

depopulation and acute inflammatory reactions in tissues. In contrast, the late reaction depends on chronic inflammatory processes, with late developing vascular insufficiency along with the development of progressive fibrosis resulting in tissue dysfunction, and which typically occur at intervals of at least six months to many years after treatment. Late reactions are important since they determine the long-term quality of life in cured cancer patients; they are highly fraction sensitive, with low radiosensitivity and α/β values, with a high repair capacity, so that the RBE values are expected to be higher on the basis of statements made the section above.

After exposures to fast neutrons (and so recoil protons), the late reactions showed by far the greatest changes in RBE with dose per fraction compared with very little change in RBE with dose per fraction in acute reacting tissues with high α/β ratios (see chapter 5). For convenience, the acute upper intestinal crypt assay became used for neutron beam inter-comparisons, as a beam quality assurance check in fast neutron and proton therapy; it is the most rapid animal tissue assay, with a high low LET α/β ratio of around 10 Gy (Gueulette *et al* 2004). Later onset and more clinically relevant effects, such as narrowing of the bowel lumen (stricture), ulceration and perforation, with α/β of around 3 Gy, are not checked by this assay. It is the predominant assay used in the *in vivo* data in Paganetti *et al* (2002).

For the above reasons, this data set (seen in figures 7.3(a) and (b)) is unlikely to show significant changes in RBE with dose per fraction. Some tissues in the same data set did include later effects, such as the lens of the eye (which represents dose-related protein opacification and not a cellular clonogenic response associated with repair) and so is not a suitable assay to make any general conclusions. A few lung assays of early and later pneumonitis (but not the later development of lung fibrosis) were included, and there are some delayed skin reactions (these show consequential late effects in animals and so the overall RBE could reflect the severity of the acute reaction). These assays, when taken together, cannot adequately represent classical late tissue reactions in a variety of human tissues (e.g. kidney, heart, CNS, gut). Also, many of the experiments on these tissues were performed at between 9–12 Gy per fraction (because such large doses were in use for proton eye treatments. This is a dose range that should reveal low RBE values compared with those expected below 2.5 Gy per fraction. It is concluded that the most appropriate experiments for defining a range of clinically relevant RBEs remain to be done for late reacting normal tissues.

Both of the Paganetti *et al* (2002) data sets with fitted lines and curves are shown in figures 7.2 and 7.3. Further analysis details can be found in Jones (2016). However, due to the data heterogeneity, fitting exercises may be unreliable. However, some broad conclusions emerge:

1. The modified *in vitro* data set can be fitted by a linear equation 1.35 – 0.02 × dose; or by an averaged LQ model fit using RBE_{max} = 1.67, RBE_{min} = 1.03 with α/β of 6 Gy.
2. The acute reacting *in vivo* data set is reasonably fitted by either a constant RBE of 1.12 or by the LQ model with RBE_{max} = 1.15 and RBE_{min} = 1.11.

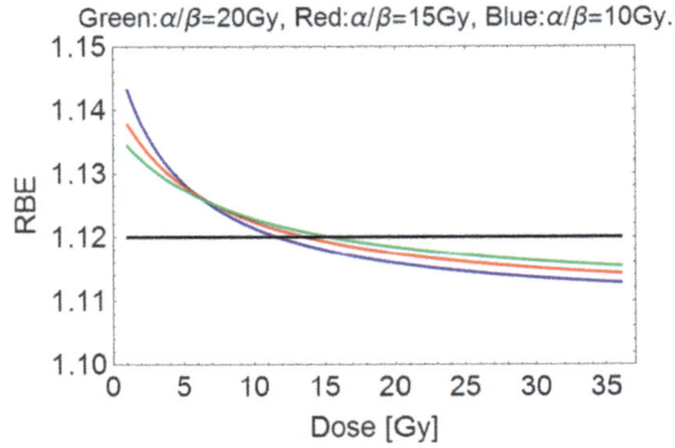

(a)

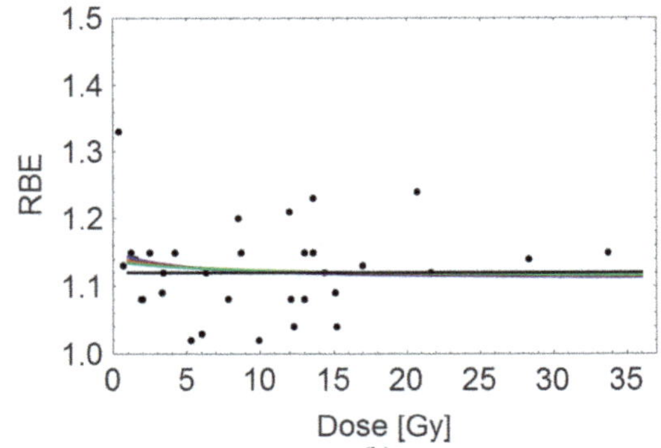

(b)

Figure 7.3. *In vivo* data (RBE > 1) showing dose per fraction and RBE plots using parameters obtained from the linear fit of 1.12 + 0.d, (shown as the grey coloured straight line), and the LQ model (for varying α/β ratios), which provides curves: (a) magnified view of the plotted models, and (b) with superimposed data over a larger scale using the same colour codes. Error bars are omitted in order to improve visual inspection, but can be seen in the original reference.

Interestingly, larger residual values were found for all attempted fits if RBE values >0 were used.

7.4 Further considerations

The existing data on proton RBEs refer to *in vitro* and acute reacting *in vivo* experiments, which are likely to underestimate the RBE in late reacting tissues, especially at lower doses per fraction. It is the latter class of tissues that contribute to long term quality of life after all forms of radiotherapy, and their tolerances must be respected in the prescription process. Paganetti *et al* (2002) indicate how difficult it would be to detect deviations in the RBE from an assumed value of 1.1; this would be especially true in experimental or clinical series where deviations in RBE may only produce clinically important changes in a small proportion of patients, depending on the degree of normal tissue sparing and the tumour dose achieved in each patient (Jones *et al* 2011b). For example, excess toxicity in the regions close to the planning target volume (PTV) would be expected if the RBE error were to override the degree of normal tissue sparing obtained. It is difficult to assess the true clinical significance, but the unexpected outcome results might occur in 5–15% of all patients. Dasu and Toma-Dasu (2013) have performed interesting clinical simulations, with significant implications for patient safety. For tissues within the PTV, the RBE error may apply to all patients, but may not be clinically significant, depending on the volume treated and the tissue functionality. For this to be detectable amid the large variation inherent in outcome studies, between 200 and 1000 patients might be required, even in the context of randomised studies (Bentzen 1994). An alternative and complimentary better approach might be to study all unexpected outcomes on an international basis and compare observed versus expected results, with a detailed analysis of LET, dose distributions, medical histories, etc in order to identify whether RBE uncertainties may be responsible. This is especially relevant to proton irradiations in the central nervous system, which have low α/β ratios of 2 Gy, with predicted RBE values >1.1 at low dose (as in chapter 9).

The new analysis also shows that further *in vitro* and *in vivo* work may be helpful to define to what extent the RBE changes with dose per fraction in many different cell and tissue types. This can be ascertained from cell survival experiments, which estimate α and β as accurately as possible, using large numbers of experiments to reduce the standard errors, in a wide panel of cells with different radiobiological characteristics. This would involve the same overall experimental plan as in Britten *et al* (2013), but with greater numbers of LET variations in order to discern if the LET–radiosensitivity turnover point position varies with incident energy, as there are suggestions in figure 7.1(a) that this might be so. Such work would contribute to solving the remaining enigma about changes in the RBE with depth in beams with different incident energies, and with different scanning modes (Calugaru *et al* 2011, Grassberger *et al* 2011). RBE–LET turnover points, where the RBE is at a maximum, are important since these can be used as a reference to scale all other RBEs (see chapters 8, 9 and 11).

In future RBE experiments, it will be essential to pursue experiments across the same broad range of dose and LET for each cell/tissue type and to analyse them independently rather than a combined analysis of all systems. This principle can be seen in the work of Sorensen *et al* (2011), where combined displays of data do not reveal what is most clearly shown for individual cell types and individual ions. Only then will the data provide useful parameters for each cell/tissue type, since the LET–RBE turnover points appear to be unique for each ion species and the RBE magnitude dependent on cell type (see chapter 8). Care must therefore be taken in the analysis of large ionic beam data sets such as in Friedrich *et al* (2013). The data analysis presented in this paper shows that analysing heterogeneous systems will arguably provide results that do not reflect well what may be occurring in individual systems. When LQ fits are so similar in a heterogeneous population, it is inevitable that LQ generated curves would show greater differences if derived from more individualised experiments.

There is now a greater appreciation that LET varies considerably in a complex treatment plan, for example when using multiple treatment field directions, or with intensity modulated pencil beams. It follows that if a constant RBE is chosen, proton therapy will not completely achieve its intended promise. This can be assessed within the formal ICRU definitions of treatment volumes (examples given in chapter 4), with potential increases in effective dose in normal tissue volumes, as well as possibilities that radiosensitive tumours (with very low predicted RBE values less than 1.1) may be under-dosed. In the absence of dose escalation, improved clinical outcomes will only be achieved if the tumour RBE exceeds that of the prescribed RBE (PRBE), and also if the PRBE is equal or less than the RBE assumed in each of the critical late reacting normal tissues. These differences should also be compared to the potential advantages of the frequently large integral dose reduction, which must influence low dose vascular and carcinogenic late effects over a much wider volume of interest. To make improvements possible, it will be necessary to allocate individual RBEs to different tissues and for different tumour types, from knowledge of the LET and dose per fraction they receive.

In comparison with the formal requirements for animal drug testing in the pharmaceutical industry, the past proton therapy experiments would be considered inadequate in terms of the range of tissues tested and the endpoints used. This is not necessarily a critique of particle therapy since this also applies to other newer techniques in radiotherapy, which are not subjected to an approved range of pre-clinical radiobiological experiments.

7.4.1 Inclusion of flexible RBEs in treatment plans

Nowadays it is possible to include a variable RBE within the treatment planning process by incorporation of suitable radiobiological models within the software. An easier and short-term solution would be to use the existing 'inbuilt' 1.1, but further reduce the tolerance of late reacting tissues. For very radiosensitive tumours such as many forms of children's cancer, it might be preferable either not to include any RBE correction, or use a lesser correction of, say, 1.05 for the tumour itself. This is because the RBE may be significantly less than 1.1, which may result in tumour under-dosage since the RBE is used to divide the photon equivalent dose to provide the physical dose of protons used. However, care must be taken not to over-dose important late reacting normal tissues where a higher RBE should be applied. The eventual gold standard might be to use dose, LET and RBE 3D maps to provide a more appropriately weighted dose, using agreed radiobiological input data. Sharing of human data transnationally or internationally or may provide greater statistical power to detect subtle but important findings when using the above techniques, for example in the collection of unexpected outcomes and their detailed analysis.

In summary, there is an urgent need for coordinated *in vitro* and *in vivo* experiments that concentrate on a realistic dose range of 1.5–6 Gy per fraction in clinically relevant tissues such as spinal cord, lung, kidney and intestine for true late effects. Also, *in vitro* experiments must be used to test a wide range of human cancer cells in resting and growth conditions, to further investigate the relationship between LET, repair capacity, radiosensitivity and dose per fraction influences on RBE.

References

Belli M *et al* 2000 Inactivation of human normal and tumour cells irradiated with low energy protons *Int. J. Radiat. Biol.* **76** 831–9

Bentzen S M 1994 Radiobiological considerations in the design of clinical trials *Radiother. Oncol.* **32** 1–11

Blomquist E, Russell K R, Stenerlöw B, Montelius A, Grusell E and Carlsson J 1993 Relative biological effectiveness of intermediate energy protons. Comparisons with 60Co gamma-radiation using two cell lines *Radiother. Oncol.* **28** 44–51

Britten R A *et al* 2013 Variations in the RBE for cell killing along the depth-dose profile of a modulated proton therapy beam *Radiat. Res.* **179** 21–8

Calugaru V, Nauraye C, Noël G, Giocanti N, Favaudon V and Mégnin-Chanet F 2011 Radiobiological characterization of two therapeutic proton beams with different initial energy spectra used at the Institut Curie Proton Therapy Center in Orsay *Int. J. Radiat. Oncol. Biol. Phys.* **81** 1136–43

Carabe-Fernandez A, Dale R G, Hopewell J W, Jones B and Paganetti H 2010 Fractionation effects in particle radiotherapy: implications for hypo-fractionation regimes *Phys. Med. Biol.* **55** 5685–700

Carabe-Fernandez A, Dale R G and Jones B 2007 The incorporation of the concept of minimum RBE (RBE_{min}) into the LQ model and the potential for improved radiobiological analysis of high-LET treatments *Int. J. Radiat. Biol.* **83** 27–39

Dale R G, Jones B and Carabe-Fernandez A 2009 Why more needs to be known about RBE effects in modern radiotherapy *J. Appl. Radiat. Isotopes* **67** 387–92

Dasu A and Toma-Dasu I 2013 Impact of variable RBE on proton therapy *Med. Phys.* **40** 01705

Friedrich T, Scholz U, Elsässer T, Durante M and Scholz M 2013 Systematic analysis of RBE and related quantities using a database of cell survival experiments with ion beam irradiation *J. Radiat. Res.* **54** 494–514

Grassberger C, Trofimov A, Lomax A and Paganetti H 2011 Variations in linear energy transfer within clinical proton therapy fields and the potential for biological treatment planning *Int. J. Radiat. Oncol. Biol. Phys.* **80** 1559–66

Gueulette J, Octave-Prignot M, De Costera B M, Wambersie A and Gregoire V 2004 Intestinal crypt regeneration in mice: a biological system for quality assurance in no-conventioanl radiation therapy *Radiother. Oncol.* **73** S148–54

Hornsey S and Field S B 1974 The RBE of cyclotron neutrons for effects on normal tissues *Eur. J. Cancer* **10** 231–4

ICRU Report 78 2010 Prescribing, Recording, and Reporting Proton-Beam Therapy (Bethesda, MD: NIH)

Jones B 2015 Towards Achieving the Full Clinical Potential of Proton Therapy by Inclusion of LET and RBE Models *Cancers (Basel)* **7** 460–80

Jones B 2016 Why RBE must be a variable and not a constant in proton therapy *Brit. J. Radiol.* **89** 1063

Jones B 2017 Clinical radiobiology of proton therapy: modeling of RBE *Acta Oncol.* DOI: 10.1080/0284186X.2017.1343496.

Jones B and Dale R G 2000 Estimation of optimum dose per fraction for high LET radiations: implications for proton radiotherapy *Int. J. Radiat. Oncol. Biol. Phys.* **48** 1549–57

Jones B, Underwood T C, Carabe-Fernandez A and Dale R G 2011a Further analysis of fast neutron relative biological effects and implications for charged particle therapy *Brit. J. Radiol.* **84** 11–8

Jones B, Underwood T C and Dale R G 2011b The potential impact of RBE uncertainty on charged particle treatment prescriptions *Brit J. Radiol.* **84** S61–9

Paganetti H 2014 Relative biological effectiveness (RBE) values for proton beam therapy. Variations as a function of biological endpoint, dose, and linear energy transfer *Phys. Med. Biol.* **59** R419–72

Paganetti H, Niemierko A, Ancukiewicz M, Gerweck L E, Goitein M, Loeffler J S and Suit H D 2002 Relative biological effectiveness (RBE) values for proton beam therapy *Int. J. Radiat. Oncol. Biol. Phys.* **53** 407–21

Sørensen B S, Overgaard J and Bassler N 2011 *In vitro* RBE-LET dependence for multiple particle types *Acta Oncol.* **50** 757–62

Tommasino F and Durante M 2015 Proton radiobiology *Cancers* **7** 353–81

Warenius H M, Britten R A and Peacock J H 1994 The relative cellular radiosensitivity of 30 human in vitro cell lines of different histological type to high LET 62.5 MeV (p-->Be+) fast neutrons and 4 MeV photons *Radiother. Oncol.* **30** 83–9

Weyrather W K, Ritter S, Scholz M and Kraft G 1999 RBE for carbon track-segment irradiation in cell lines of differing repair capacity *Int. J. Radiat. Biol.* **75** 1357–64

IOP Publishing

Practical Radiobiology for Proton Therapy Planning

Bleddyn Jones

Chapter 8

A general RBE simple efficiency model for protons and light ions

8.1 Introduction

Positively charged particle therapy is associated with increased linear energy transfer (LET), creating clustered and complex DNA damage, which is less repairable and so causes enhanced biological effects. This changes normal tissue radio-tolerances, as well as tumour control probabilities, with accompanying changes in fractionation sensitivity. LET can be expressed in different ways, either as the mean or the dose-averaged LET, but there are other variants (see chapter 1). Comparative changes in bio-effects are quantified by the RBE concept, defined and measured as the ratio of dose of a low LET radiation divided by the control high LET dose required for the same biological effect. RBE depends on the complexities of how radiation of different qualities interacts with different biological systems due to the following.

(1) The LET depends on the particle energy, atomic charge, position along track (or tissue depth) and the mixture of Bragg peak or non-Bragg peak regions over a volume of interest (Newhauser and Zhang 2015).
(2) The increased local complexity, or clustering, of DNA damage with increasing LET (Anderson *et al* 2007).
(3) The increasing difficulty in repairing the more clustered damage, resulting in increased radiosensitivity and reduced fractionation sensitivity (see chapter 2).

Some authors (e.g. Hall and Gaccia 2011) have tried to unify LET effects by claiming that the LET–RBE (relative biological effect) turnover point is common for all ion species at around 100–120 keV μm^{-1}, and that this value is somehow determined by the average DNA strand thickness. This is not the case on detailed examination of data sets for each ion species, for example those displayed by Sørensen *et al* (2011), where there appears to be a unique value for the RBE turnover point for each ion species. Furthermore, it can be seen that the turnover point LET value increases with the atomic number (Z) and that the maximum RBE achieved by any ion is strikingly similar.

Some authors have developed relatively simple LET–RBE models for protons (Wilkens and Oelfke 2004, Carabe *et al* 2012, Jones 2015A). For ion beams, there are several complex formulations that tentatively describe the relationship between LET and the RBE (Katz *et al* 1971, Hawkins 2003, Kase *et al* 2008, Elsässer *et al* 2008, Inaniwa *et al* 2013, Grün *et al* 2015), each with varying degrees of success, and have been used for clinical applications. These ion beam models are based on the fundamental interactions of particle physics with matter and contain multiple assumptions and input requirements, such as knowledge of particle trajectories relative to cells, cross-sectional probabilities, the relative proportion of cell nucleus to cell volume for each cell, critical biological sub-volumes, repair capacities, extrapolations with dose, etc. They all utilize long mathematical constructs, which can be daunting to less mathematically gifted individuals. Whereas it is satisfying to build exploratory theoretical models in such a way, it is impossible to know these exact conditions within a real cancer and surrounding normal tissues. These various approaches have been used to predict ion beam RBE values for variable LET values for the irradiation of specific cell types (usually the V-79 cell derived from Chinese hamsters), but with mixed results, although they are used routinely in clinical practice for carbon ion treatment planning. Only some authors have attempted an approach for normalizing the RBE differences between different ions, as in the work of Katz *et al* (1971), who developed the useful concept of the Katz radius of $Z^{*2}/(v/c)$, (where v is the particle velocity and c the speed of light; the ratio is usually expressed as β in physics although in the context of radiobiology it should not be confused with the second linear-quadratic (LQ) model radiosensitivity coefficient). Since the radius is smallest for proton with $Z = 1$, than for all other ions, the ionisations are closer for protons and so capable of producing increases in RBE at relatively low LET values. As the Z values increase, the LET value of the maximum RBE with LET occur at increasingly higher LET values, as discussed below and with further data analysis given in chapter 12.

What do we know with certainty about LET and the RBE? Measured relationships between LET and the RBE generally show increases with LET until a maximum value is achieved, followed by a decrease to RBE values just above unity. Also, there are important basic findings that models must incorporate in order to adequately describe how RBE changes with LET. These facts can be found in the reports of Barendsen (1968), Weyrather *et al* (1999) and Sorensen *et al* (2011). They are as follows.

- The initial slope of RBE with LET is linear when plotted on linear scales.
- The LET value (LET_U), which confers the maximum cell killing efficiency (at the turnover point), increases nonlinearly with the nuclear charge of a particle (the Z number), which denotes the electrostatic positive charge of the particle nucleus. LET_U values increase with Z, but smaller LET_U increments are apparent with increasing Z, which suggests a saturation effect. Ions with the smallest Z values are consequently more efficient in increasing RBE per unit increase in LET, possibly because the energy released is more locally absorbed than is the case for higher Z ions with larger event sizes and more energetic gamma emissions.

- The magnitude of the RBE is not only dependent on the particle type (or Z), but also depends on LET, dose (and the surviving fraction of cells), and the cell type (and its ability to repair radiation damage).
- The magnitude of the RBE for any ion also depends on cell type, with cells that are intrinsically more radiosensitive (to very low LET radiations) having lower RBEs than their more radioresistant counterparts.
- The RBE increases with dose reduction (when the cell-surviving fraction is reduced), but the LET_U (turnover point position) remains constant. Thus, the overall LET–RBE symmetry is preserved despite RBE decreasing with dose.
- In terms of the LQ model of radiation effects, the α value increases with LET to a far greater extent than β (Hawkins 2009, Jones 2010, Jones *et al* 2011, Jones 2015b). The relative increase in α with LET is greatest for cells/tissues that have the lowest, most radioresistant, low LET α_L values, with smaller increases in α for systems that have the most radiosensitive, highest intrinsic α_L values, as will be shown later. A similar effect is assumed for β, and has been identified in some experiments (Jones 2015b, Weyrather *et al* 1999).
- These experimental findings apply to well-oxygenated cells, but are modified in radiologically hypoxic conditions (Furusawa *et al* 2000), probably since the α parameter related cell kill is not so influenced by oxygen as is the β parameter related cell kill (Jones *et al* 2007).

This chapter describes the development of a relatively simple LET model using efficiency–inefficiency considerations to estimate RBE relationships, and their modification with dose, Z, and the two low LET intrinsic radiosensitivities α_L and β_L. This model requires fewer input parameters and assumptions than do the far more complex models already referred to above. Such a model could be used to complement the other systems; in this sense, two predictions using different models are probably more reliable if they are in reasonably close agreement.

8.2 The experimental data and its limitations

There are relatively few published experiments that provide a reasonable estimate of LET turnover positions (LET_U) for clinically used particles. These include the data sets of Belli et *al* (2000), Barendsen (1968), Furusawa *et al* (2000), and Weyrather *et al* (1999). These experiments were not designed to accurately determine LET_U, but to show overall phenomena and to determine the range of RBE values. Inevitably, overall accuracy is further undermined by biological variation, use of different cellular assays in various laboratories, use of different LET interpretations and measurements over wide ranges with consequent use of a logarithmic scaled abscissa, which masks the uniform initial linear slope of the relationship. To obtain the best available estimate of LET_U, only the most unequivocal examples of maximum radiosensitivities, or the RBE, over a small range of LET near to LET_U, were used. Data where LET_U could not be determined to reasonable accuracy, as in some of the HRG cellular data of Furusawa *et al* (2000) and in some carbon ion experiments, were excluded. The LET_U values (keV μm^{-1}) so obtained were 30.5 (protons), 103.4 (helium), 208 (carbon), and 233 (neon). Although data exist for heavier ions such as silicon and argon (Friedrich *et al* 2013), these do not provide sufficiently accurate estimates of LET_U.

As a test of model fitting for each unique ion, the following experiments are used:

1. Todd (1967), deuterons, helium, lithium, boron, carbon, oxygen, nitrogen, neon, and argon ions,
2. Barendsen (1968), deuterons and helium ions, and
3. Weyrather *et al* (1999), using carbon ions.

The highest α radiosensitivity obtained (α_U), in the region of LET_U, for each ion species, was plotted against the low LET (control) α_L value from the same data. These values are shown in later graphical plots, and include variation due to the LET_U position uncertainty. The accuracy of the β radiosensitivity parameter is less easy to determine for high LET radiations (compared with low LET radiations), for reasons discussed elsewhere (Jones 2015b). It is well established that β increases to a lesser extent than α with LET, but are their LET turnover points the same? In order to maintain the observed constant position of the maximum RBE occurring at LET_U, and this being constant at any dose (or surviving fraction), then in order to maintain the overall symmetry of the LET–RBE relationship, both α and β must follow symmetrical functions, each of which rises to a maximum value at LET_U. Otherwise, the overall symmetry of the LET–RBE curves with increasing dose would be broken. For instance, if LET_U was different for α and β, the LET_U would be observed to change with dose, which is not the case. Figure 8.1 shows the overall symmetry of the LET–RBE relationship for different doses together with the phase space limits given by the plots of RBE_{max} (or α_H/α_L)—at near zero dose and RBE_{min} (or $\sqrt{\beta_H}/\sqrt{\beta_L}$) at very high dose. The method for constructing these plots is described below, but it is useful to observe this effect here.

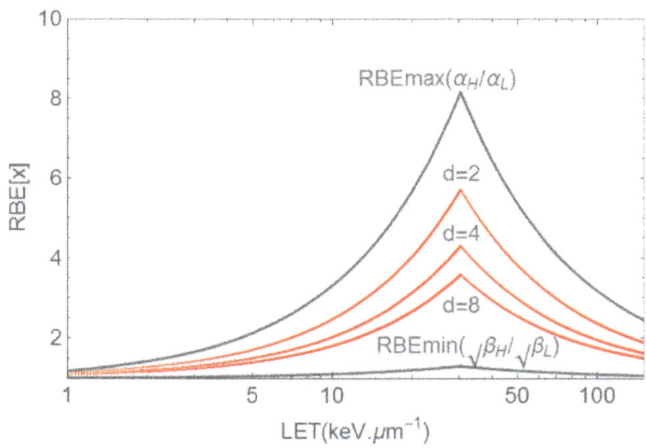

Figure 8.1. Theoretical plots of RBE_{max}, RBE_{min} and RBE values at doses of 2, 4 and 8 Gy with low LET α parameter of 0.1 Gy^{-1} and α/β ratio of 3 Gy, consistent with a normal tissue late effect. The overall symmetry can be seen for each plot. Thus the LET_U value here is typical of protons, around 30 keV μm^{-1}, for each of α, β and RBE, irrespective of dose.

8.3 Description of the Z-specific model

8.3.1 The relationship between Z and LET_U

The position of the turnover point appears to be unique for each Z value. It is apparent from publications quoted above (Barendsen 1968, Furusawa *et al* 2012, Weyrather *et al* 1999, Belli *et al* 2000). that LET_U increases with Z, but the effect appears to saturate (i.e. further increases in LET have diminishing returns as far as the LET_U value is concerned). The Betha–Bloch equation for estimating the rate of energy loss with distance (x) traversed (dE/dx), which represents LET, contains a Z^2 term in the numerator usually reflecting the charge of a fully electron stripped ion or proton. Larger Z values will also be associated with larger mass numbers and greater momentum with larger event volumes due to more complex nuclear collisions and energetic γ-ray emissions. Beyond the necessary critical dimension (be this radial or linear as a surrogate), the biological killing efficiency will not increase if the event size becomes too large and physically beyond the individual chromosome. So, a saturation effect is to be expected. The smallest values of $Z = 1$ for a proton effectively reduces dE/dx, but the proton LET_U is only 30.5 keV μm^{-1}, suggesting that lighter charged particles exert more localized effects (caused by short range low energy secondary electrons). In this respect, the proton is more efficient at causing an increment in RBE with LET. (However, proton LET values are quite small. For example, averaged LETs of only 1–8 keV μm^{-1} occur in radiological voxels during typical clinical exposures (Grassberger *et al* 2011) and when using scanned proton beams may cause RBEs as high as 1.8 or more (Jones 2015a).)

The application of a simple differential equation can represent this process. Let us assume that Z is a continuous variable and if the initial rate of change in LET_U with Z is S and that this value then decreases in proportion to LET_U itself, representing a saturation effect controlled by the constant k, so that

$$\frac{dLET_U}{dZ} = S - k\ LET_U \tag{8.1}$$

which by integration of both sides and rearrangement leads to

$$LET_U = S/k(1 - \exp[-k(Z)]), \tag{8.2}$$

where S/k represents the maximum possible value of LET_U.

Equation (8.2) can be normalized to the proton ($Z = 1$) LET_U of 30.5 keV μm^{-1} found by Belli *et al* (2000), so that for any Z a term $Z - 1$ is used such that:

$$LET_U = 30.5 + S/k(1 - \exp[-k(Z - 1)]). \tag{8.3}$$

This equation is used for data fitting purposes.

8.3.2 Changes in the radiosensitivities with LET

By increasing LET gradually, from the control low LET value of say clinical 4–6 MV photons (x-rays), we obtain small increases in the probability of additional lethal chromosomal breaks. The energy deposition becomes maximally efficient (let this be represented by 100% efficiency for normalization purposes), and at higher LET values beyond LET_U, the efficiency is reduced below 100% due to excess local energy deposition, which does not result in more effectiveness.

The separate relationship between α_L (the low LET control α value) and α_U (the value of α at the turnover point where LET = LET_U) also exhibits saturation effects. In other words, the increment in α with LET show diminishing returns, since the lowermost α_L values have the highest gain in α. This effect is found with fast neutrons and with charged particle data, as shown in section 8.4.

For an initial slope of A and a rate constant j, the rate of change of α_U with α_L will fall in proportion to α_U, so that

$$\frac{d\alpha_U}{d\alpha_L} = A - j\alpha_U, \tag{8.4}$$

which leads after integration to:

$$\alpha_U = A/j(1 - \exp[-j\alpha_L]). \tag{8.5}$$

The β parameter can either be modelled in a similar way, but with smaller overall changes, or to simplify matters for tentative modelling purposes, it could be assumed to be invariant with LET, but only at low doses where β-related cell kill is small and where low LET α/β ratios are high. It is preferable to use β-modification for low α/β value tumours and late reacting normal tissues. The data of Weyrather *et al* (1999) show that β values rises from a control value of 0.026 Gy^{-2} to a maximum of 0.044 Gy^{-2} in V-79 cells, and likewise from 0.02 to 0.042 Gy^{-2} for the CHO cells (one value of 0.192 Gy^{-2} in this data set must be erroneous due to the fitting program or other assay-related issues. So, β appears to increase by up to a factor of around two, which is small compared to the maximum increments in α with LET of around 10. There is more abundant data for 64 MV fast neutrons, where β undoubtedly increases (Jones 2010). Although such neutron experiments will probably underestimate the maximum possible rise in α and β, since the neutron LET spectrum (and its average value) may not necessarily be close to the LET_U for an ionic beam. Nevertheless, further analysis of these data, which compare neutrons with megavoltage x-rays show fits of $\beta_{neu} = 1.54\, \beta_x$-ray or $\beta_{neu}\, 0.097$ (1 – exp (23.6 β_x-ray), as will be shown below). The experimental variation in such data is considerable and the two fitted equations were obtained after elimination of repair deficient cells (where $\alpha > 0.6$ Gy^{-1}), or where neutron β values close to zero, or if the increment in β exceeded that in α (suggesting experimental artefact). It should be noted that α_U and β_U values will be higher than the maximum values obtained for fast (64 MV) neutrons, and so the neutron data cannot be used directly to determine RBE changes for ion beam data.

A similar ‘saturation’ function, is used to link β_L with β_U, as given elsewhere (8):

$$\beta_U = (R/u)(1 - \exp[-u\beta_L]) \tag{8.6}$$

where $R = 2.5$ and $u = 25$, which provides a modest increase in b, and is compatible with the limited data discussed already, and with a maximum ceiling value of 0.1 Gy^{-2} for β_U.

8.3.3 Obtaining α_H and β_H values

In simple mathematical terms, a discontinuous or biphasic (efficiency followed by inefficiency) model can be used, where for LET values up to that of LET_U increasing efficiency is represented as a linear simple proportional relationship, as used by Wilkens and Oelfke (2004) for protons, and where the α value at any LET higher than the control and lower than the turnover value will be

$$\alpha_H = \alpha_L + \frac{LET_x - LET_C}{LET_U - LETc}(\alpha_U - \alpha_L) \tag{8.7}$$

where α_H is the α value at any particular LET value (LET_x) between the control and ultimate value of LET_x (which represents any LET value between the control value of LET_C (where α is α_L)] and LET_U, where the maximum α of α_U occurs.

For the initial linear portion of the relationship, there will be a uniform gradient of

$$\frac{(\alpha_U - \alpha_L)}{(LET_U - LET_C)} \tag{8.8}$$

between the values of LET_C and LET_U, which fulfils the requirement for linearity in this LET range.

It follows that, for example, if LET_C and LET_U are 1.2 and 120 KeV μm^{-1} respectively, with α_L and α_U of say 0.3 and 1.3 Gy^{-1}, respectively, then for a LET_x value of 60 the process is only $(1.3 - 0.3)/(120 - 1.2) \times (60 - 1.2)$, which is close to being 50% efficient, and for a LET_x of 90, the efficiency will be $(1.3 - 0.3)/(120 - 1.2) \times (90–1.2)$, which is close to 75% efficiency.

In this way, the efficiency of cell kill per unit dose will increase linearly with LET, leading up to maximum efficiency (defined as 100%) at LET_U.

For values of LET beyond the turnover point (where LET > LET_U), the additional energy transferred does not contribute to extra lethality, but is wasted. That is, the excess energy ($LET_x - LET_U$) beyond the optimal released energy is wasted. Consequently, inefficiency, expressed in energy terms by energy wasted/energy released, or $(LET_x - LET_U)/(LET_U - LET_C)$ increases with LET. To express this in terms of efficiency, the engineering relationship of % efficiency = 100 − % inefficiency is used, and the α_H value is then scaled between α_U and α_L.

Accordingly, the equation for α_H for LET > LET_U then changes to be:

$$a_H = a_L + \left(1 - \frac{LET_X - LET_U}{LET_U - LET_C}\right)(\alpha_U - \alpha_L) \tag{8.9}$$

which effectively expresses a progressive reduction in α with further increasing LET. In this way, if LET_x is 180 and LET_U is 120, the value of α_U at the turnover point of 100% efficiency will fall to 1 − (180 − 120)/180, which provides around 67%

efficiency. For a LET_x of 240, we obtain 1 – (240 – 120)/240, which is 50% efficient. These efficiencies are of course relative to a normalized value of 100% at the turnover point.

Similar equations are used to provide β_H by proportionate scaling between β_L and β_U. These are obtained by simply replacing α_L, α_H, and α_U by β_L, β_H, and β_U, respectively in equations (8.7) and (8.9), which are not reproduced here.

The estimated RBE is then obtained using the iso-effect equations given below in section 8.3.5.

The above method can be used to tentatively estimate RBE for any cellular system based on generic assumptions based only on the input value of the low LET radiosensitivity parameters. Alternatively, if there is available data sufficient to estimate the slope of the radiosensitivity parameters with LET then it is easy to estimate the RBE for any LET value within the range of the experiments, as described in the next section.

8.3.4 An alternative method that does not use LET_U, but uses the slope of the radiosensitivity or measured RBE increments with increasing LET (up to the turnover point)

For RBE measurements at specified doses or bio-effects (such as surviving fraction), simple linear fitting should be sufficient, although in some instances a nonlinear fit may be required if the relationship deviates from linearity. One cannot easily then predict the RBE at different dose or effect levels in the absence of radiosensitivity information.

Where detailed radiosensitivity measurements have been made, it is easy to estimate the slopes of α and β increasing with LET (noting that some cell lines show little or no increase in β). If the slopes are, respectively, S_α and S_β for each radiosensitivity parameter, then

$$\alpha_H = \alpha_L + S_\alpha(LET_x - LET_C),$$

and

$$\beta_H = \beta_L + S_\beta(LET_x - LET_C),$$

then the RBEs can be estimated at any dose by applying the equations in the following section.

8.3.5 The RBE at any specified dose per fraction

The reduction in RBE with reduced surviving fraction and increasing dose is obtained by the solution of the following iso-effect equation for high and low LET radiations at a dose d_L and d_H, for low and high LET respectively:

$$\alpha_L d_L + \beta_L d_L^2 = \alpha_H d_H + \beta_H d_H^2. \tag{8.10}$$

The solution for d_L is then divided by d_H to provide the RBE, as shown in other publications (Jones 2015a, 2015b).

For clinical iso-effect calculations, the solution of the following biological effective dose (BED) equations are used for the low and high LET:

$$n\, d_L \left(1 + \frac{d_L}{\left(\frac{\alpha}{\beta}\right)_L} \right) = m\, d_H \left(\mathrm{RBE}_{max} + \frac{\mathrm{RBE}_{min}^2 d_H}{\left(\frac{\alpha}{\beta}\right)_L} \right) \tag{8.11}$$

where n and m are the respective numbers of fractions for the low and high LET.

The RBE parameters are replaced by LET (and the new parameters given in the sequence of equations described above) and then solved for d_H. Total doses to provide the same BED can then be calculated for different numbers of fractions.

8.4 The graphical results

8.4.1 Radiosensitivity data

The relationship between Z and LET_U shown in figure 8.2, using pooled data for proton, helium, carbon, and neon ions (Belli *et al* 2000, Barendsen 1968, Weyrather *et al* 1999, Furusawa *et al* 2000), were fitted by equation (8.3).

The Clatterbridge fast neutron data (Warenius *et al* 1994) that show the relationship between α_L (for values up to 0.8 Gy^{-1}) and α_H, and between β_L and β_H, are shown in figures 8.3(a) and (b), respectively. In each case, the linear and non-linear fits are not significantly different ($p > 0.05$), although the residuals are smallest for the nonlinear equations, which also have the advantage of not extrapolating to infinitely high radiosensitivity values.

The relationship between α_L and α_U for various ions are shown in figure 8.4, fitted to data from the literature (with data where negative β values obtained excluded). The fitted equation is shown in the figure, but also with a least squares

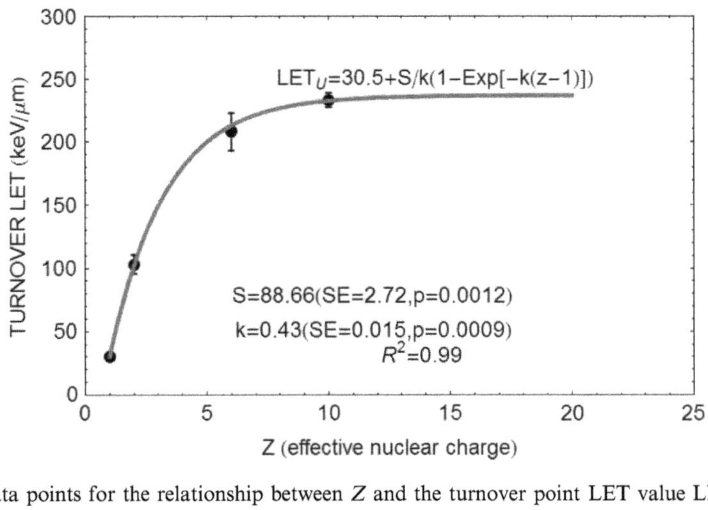

Figure 8.2. Data points for the relationship between Z and the turnover point LET value LET_U with fitted parameter values based on equation (8.3). Reproduced from Jones (2015b).

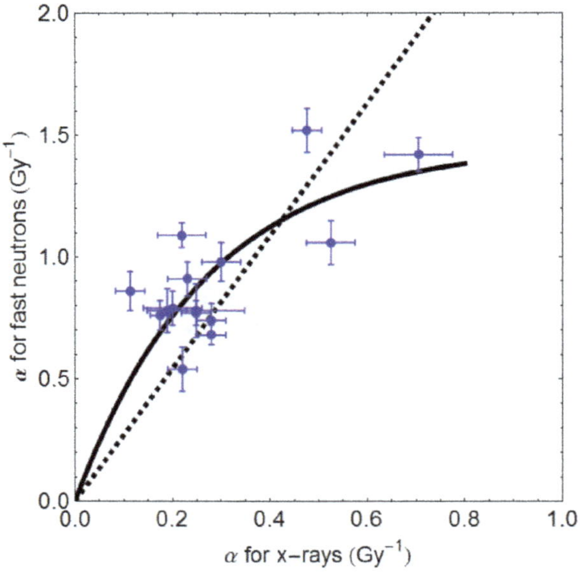

(a)

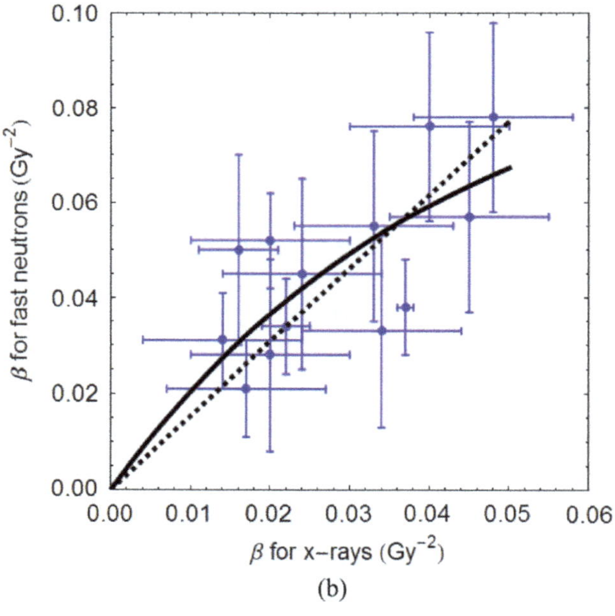

(b)

Figure 8.3. 64 MV fast neutron relationships between low and high LET radiosensitivity parameters. Linear no-intercept, and nonlinear least-squares fits are, respectively: (a) $\alpha_H = 2.72\alpha_L$, and $\alpha_H = 5.37/3.68(1\text{-}e^{-3.68\alpha L})$; (b) $\alpha_H = 1.57\alpha_L$ and $2.29/23.57(1\text{-}e^{-23.57\ \alpha L})$. Reproduced from Jones (2015b).

fit for a linear no-intercept relationship of $\alpha_U = 6.47\ \alpha_L$ ($p < 0.001$, $R^2 = 0.899$) for α_L values less than 0.35 Gy^{-1}, the more radioresistant part of the radiosensitivity spectrum.

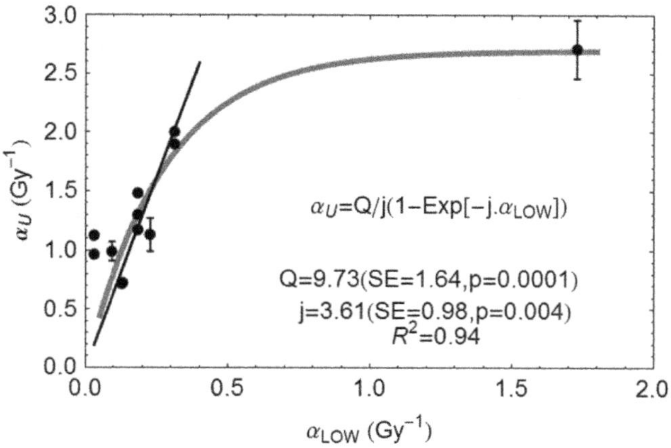

Figure 8.4. Ion beam relationships between radiosensitivity parameters at low and high LET at the turnover point (α_H is here α_U) and fitted by the parameters shown. Error bars are not available for all data used. Reproduced from Jones (2015b).

8.4.2 Fits to experimental RBE data sets

The model is superimposed to the experimental data sets, using different cell lines, of Barendsen (1968) (figure 8.5) and Weyrather *et al* (1999) (figures 8.6(a) and (b)) and Todd (1967) (figures 8.7 and 8.8).

The data of Barendsen used mono-energetic deuterium or helium (alpha) particles in one human cell type, with highly symmetrical curves that turnover at around 110 keV μm^{-1}. In this case (see figure 8.5), the model fits the data reasonably well at all levels of surviving fraction. However, since this data set exists as plotted graphical surviving fraction results without access to the original data, there is inevitable uncertainty in assessing the low and high LET α and β values, which make the RBE determination even more difficult. The plot was obtained by assessment using $\alpha_L = 0.16$, $\alpha_U = 1.31$, $\beta_L = 0.046$ and $\beta_U = 0.15$ obtained by crude measurements of survival curves and RBE plots, each on logarithmic and linear scales, drawn by artists, and which contain displacements of many data points for convenience of display, but the data set is better fitted by $\alpha_L = 0.15$, $\alpha_U = 1.35$, $\beta_L = 0.03$ and $\beta_U = 0.08$, as shown in figure 8.5. The Barendsen data set suffers from retrospective inaccuracies in estimating parameters from diagrams in publications rather than use of the raw data, but the graphic shows the sensitivity of the model to the input parameters.

The critical dependency of each RBE limit on the ratio of α and β at low and high LET respectively, demonstrates the importance of obtaining the most accurate possible data, rather than depending on published material that does not contain precise surviving fraction outcomes. The Barendsen data set also suggests a higher value of LET_U for alpha particles than obtained above using the formula based on Z in pooled data, at around 127 instead of 103 keV μm^{-1}; also the α_U is predicted to be 1.18 Gy^{-1} by equation (8.3). This illustrates the uniqueness of each data set and the distorting effect of pooling of data from different laboratories using different cell systems, etc.

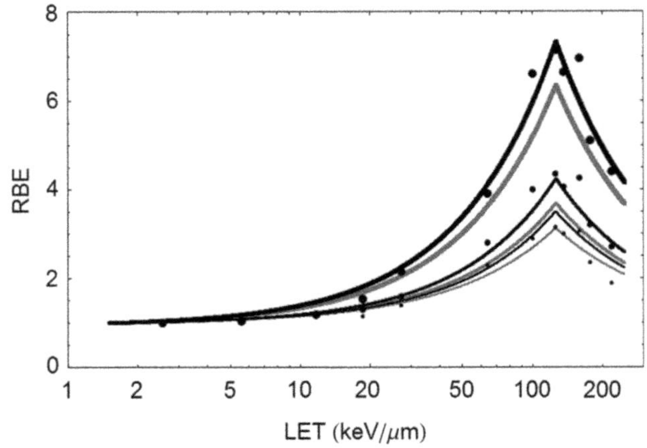

Figure 8.5. Mono-energetic alpha particle data of Barendsen using the proposed model to provide fit lines, with black indicating use of parameters $\alpha_L = 0.16$ Gy^{-1}, $\alpha_U = 1.31$ Gy^{-1}, $\beta_L = 0.046$ Gy^{-2} and $\beta_U = 0.15$ Gy^{-2}, and grey using parameters $\alpha_L = 0.15$ Gy^{-1}, $\alpha_U = 1.35$ Gy^{-1}, $\beta_L = 0.03$ Gy^{-2} and $\beta_U = 0.08$ Gy^{-2}. The thickest lines are for 50% survival, medium lines for 10% survival and thinnest lines for 5% survival. Reproduced from Jones (2015b).

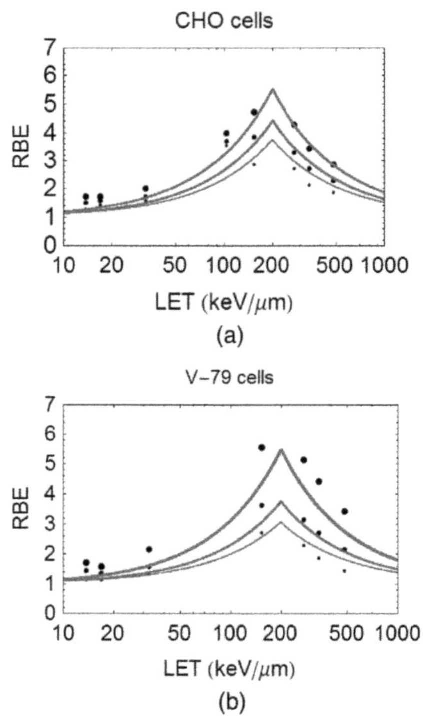

Figure 8.6. Model fitted data of Weyrether *et al* (1999) for C ions for three different cell lines and doses, coded in the same way as for figure 8.4 with respect to line thickness and surviving fraction, (a) for CHO cells, and (b) for V79 cells. Reproduced from Jones (2015b).

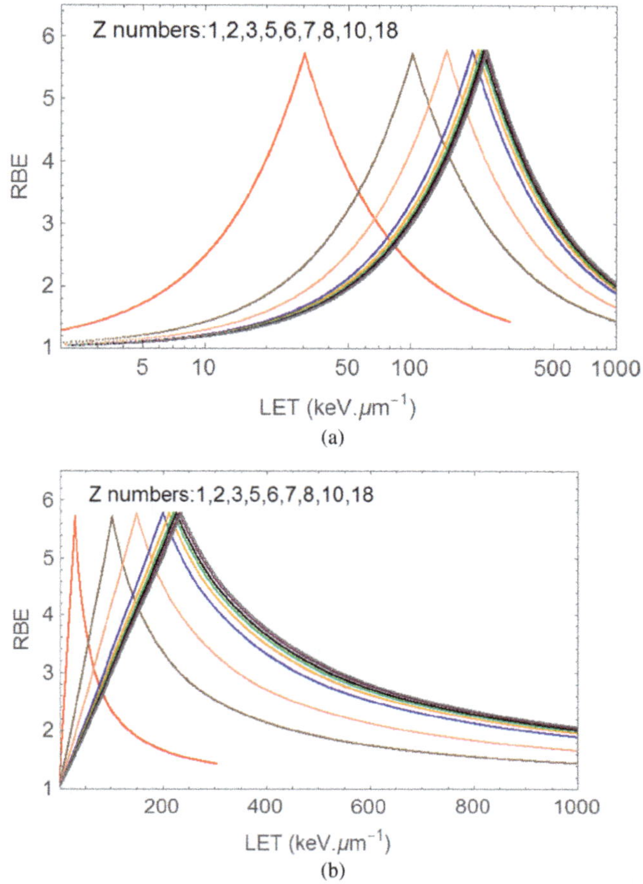

Figure 8.7. Graphical displays of RBE and LET: (a) with the familiar log scaling on abscissa and (b) with linear scaling, each with unique turnover point positions for multi-ion data of Todd (1967), assuming $\alpha = 0.14$ Gy^{-1} and $\beta = 0.05$ Gy^{-2} for a dose of 1.5 Gy. From left to right the ionic elements are shown as follows, with colour code and LET_U (rounded to nearest integer for values over 100) in parentheses: deuterium (red, 30.5), helium (brown, 103), lithium (pink, 150), boron (blue, 200), carbon (orange, 213), nitrogen (green, 221), oxygen (Black, 227), neon (purple, 232) and argon (grey, 237); the respective Z numbers are indicated on the graphic. Reproduced from Jones (2015b).

The important carbon ion data of Weyrather *et al* (1999), from GSI, which covers a broader range of LET values, shows an apparently constant turnover point for different cell types and surviving fractions (figure 8.6). The data are published with the LQ radiosensitivities, although the ions have a small variation in their LET spectrum (with a maximum spread of less than 5% for the highest LET values, which reduces further with decreasing LET). So, it is unlikely that energy and LET spread contribute to the deviations from the modelled curves seen at lower LET values. The RBE values found at low LET values seem higher than expected, possibly due to biological sample variation, especially since irradiations were performed using two different accelerator systems (for LET values above and below 100 keV μm^{-1}) in different laboratories and presumably at different times. These data, although very

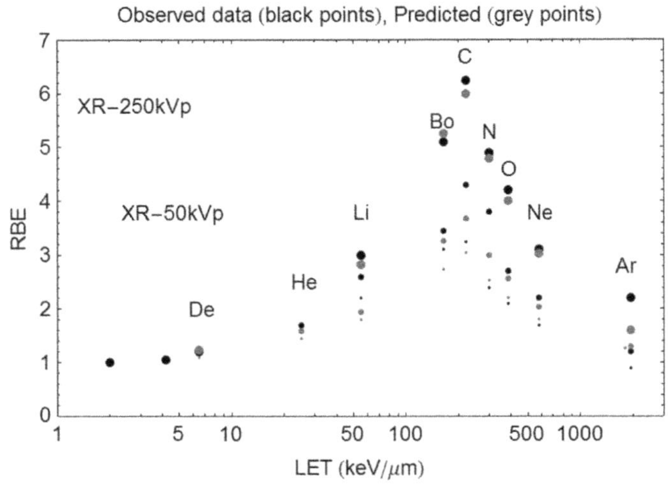

Figure 8.8. Model predictions shown with data of Todd using the Z-based model to determine LET_U and α_U, β_U respectively, with LQ model correction for dose. The largest sized points are for SF = 50%, the intermediate sized points are for SF = 10% and the smallest sized points are for SF = 1%. The ions used and their LET (keV μm^{-1}) values are given in parentheses. Deuterium (6.5), helium (25), lithium (55), boron (165), carbon (220), nitrogen (300), oxygen (385), neon (580) and argon (1940). Observed RBE data are printed as black points, with estimated RBE values as grey points. Starting on the left-hand side the first two black points are for 250 and 50 kVp x-rays respectively, followed by deuterons, etc. Reproduced from Jones (2015b).

informative, inevitably contain greater heterogeneity than the data of Barendsen, and the data are less well fitted. Another more stochastic approach is to use a Poisson function, which will be presented in a further publication.

In the case of Todd's multi-ion data, a range of different mono-energetic ions were used (protons, deuterium, helium, lithium, boron, carbon, nitrogen, oxygen, neon and argon), which implies that there will be at least nine different curves, one for each Z value, and each with unique turnover points. Such heterogeneous data were fitted surprisingly well by allocating a unique turnover point for each ion species, before estimation of the RBE, as shown in figure 8.7, followed by the RBE estimations for each ionic species in figure 8.8.

8.4.3 Extensions to clinical radiobiology

It is possible to tentatively assess changes in total dose required for different fractionation schedules using protons, helium, and carbon ions, as shown in figures 8.9(a)–(c). The variations in LET are representative of the wide expected clinical ranges for non-Bragg-peak regions, spread out Bragg peaks of different sizes, and for scanned beams. It should be noted that the changes in total dose required with number of fractions (and consequently dose per fraction) are remarkably similar for the respective LET ranges used. This indicates the importance of LET mapping as well as dose mapping in the clinic, since RBEs and consequently changes in total dose with fractionation can be the same for a wide range of ions, as determined by their Z value and LET.

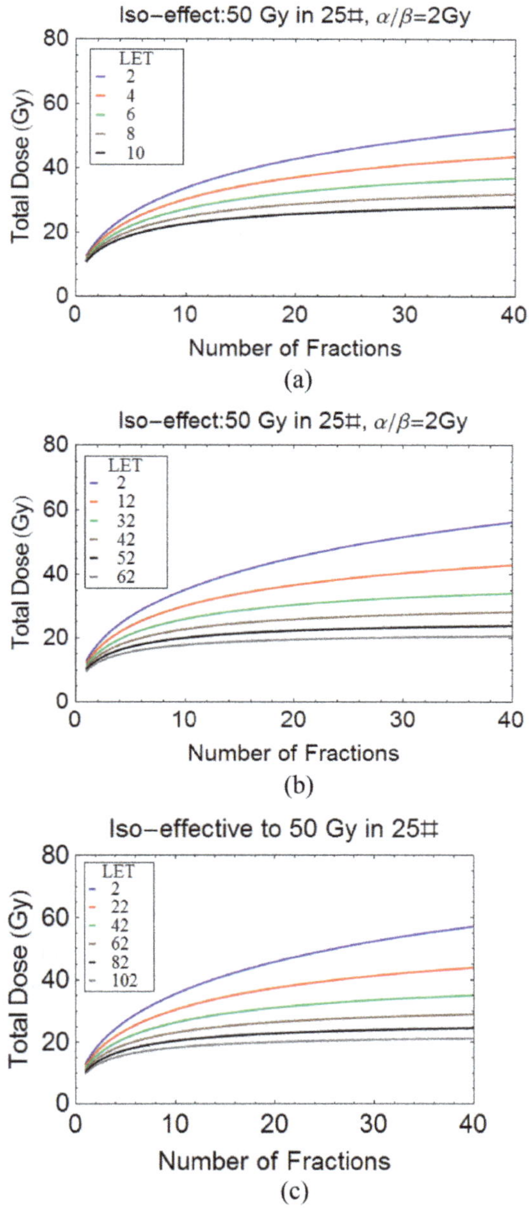

Figure 8.9. Plots of total iso-effective dose versus number of fractions for the given iso-effect and α/β ratio: (a) protons, (b) helium ions and (c) carbon ions. Reproduced from Jones (2015b).

8.5 Conclusions and what remains to be done

Simple differential equations that model saturation effects are commonly used in the physical sciences and in biology, with notable examples in pharmacokinetics. Saturation in the radiation context applies to the relationship between the effective radiation event size and the bio-target. Maximum efficiency represents the maximum cell killing effect caused by locally absorbed energy, which differs from the energy released, some of which may be wasted by causing more local damage than is necessary to cause lethality, or is dissipated over a wider than necessary critical volume.

The new model offers a relatively simpler semi-empirical mathematical method for assessing changes in RBE with LET than has previously been available and provides a second order approximation. It can be more easily understood and used by clinicians, biologists, and others, without recourse to more complex mathematics. Also, the two saturation-based assumptions made, in comparison, are fewer than the assumptions required in other RBE models. For highly controlled and relatively homogenous data sets, this deterministic approach provides reasonable estimates of RBE by obtaining RBE_{max} and RBE_{min} from the incremental values of α and β at increasing LET. The model depends on the assumptions of the LQ model where α and β are high level parameters, being ultimate coefficients of radiation induced cell death, rather than basic components of radiation effects such as DNA strand breaks, etc.

The model is not intended to supplant existing models of the RBE, but to be complementary. It would be highly advantageous in clinical practice if more than one model could be used, with clinical decisions allowed to proceed if at least two are in reasonable agreement. Thus, the LEM, MKM, and variants of the Katz models should continue to be used, and compared with the new model.

Improved input data would undoubtedly further improve the accuracy of the model. Rather than attempt to fit historical data, which are limited in terms of accurate determination of 'maximum efficiency' turnover points, it would be better to conduct rigorous experiments to test hypotheses connected with the above models, such as the relationship of the initial slope to more precise estimates of the turnover point position (LET_U) in different ions. This also requires a further stochastic interpretation necessary to match a range of LET values as would be encountered in many clinical beams. There is ample scope for research in this respect.

Some authors have emphasized the inverse association between low LET α/β and the final RBE (Carabe *et al* 2012, Hawkins 2003, Elsässer *et al* 2008). This follows since α/β represents cellular repair capacity and the intrinsic radiosensitivities, but is more valid at low doses. From the definitions of RBE_{max} and RBE_{min} it is easy to show that the former will be inversely related to $(\alpha/\beta)_L$, but the latter directly proportional to the square root of $(\alpha/\beta)_L$ (Jones *et al* 2011), as shown in chapter 8.5. The former assumption can be used for low dose per fraction treatments, where RBE_{max} dominates the RBE. The need to include changes in β with LET is necessary for estimations of RBE at higher doses, and where α/β is small as in human late tissue effects. The new model also preserves the overall symmetry of the curves at increasing dose. Accurate estimation of β from cell survival curves, especially when α values are large, is notoriously difficult to achieve. Our knowledge of how β changes with increasing LET is less well documented than for the larger and easier to measure changes in α with increasing LET. Only by meticulously conducted large scaled experiments, with greater than usual numbers of cell survival experiments, can these parameters be estimated to greater and sufficient accuracy.

Since neutrons are uncharged, they do not fall easily into this model, although the main products of neutron interactions such as recoil protons and other ions do, such that a spectrum of LET values will result, which, in principle, could be translated into the RBE using the modelling described in this book. A crude estimate could be obtained from the specific or representative neutron LET (even if chosen from a spectrum) and applying $Z = 1$ in the model described.

There is considerable scope for the application of simpler RBE predictive models. Ideally prospective experiments should be performed with specific attention to LET–RBE turnover point position for different ions, the initial slope of the increment in RBE and the maximum value of α and β relative to their low LET values. These values need to be determined for extensive *in vitro* libraries of human cell lines and, if confirmed, extended to more complex *in vivo* experiments.

The next chapter will use the same model but for the specific application of protons.

References

Anderson R M, Stevens D L, Sumption N D, Townsend K M, Goodhead D T and Hill M A 2007 Effect of linear energy transfer (LET) on the complexity of alpha-particle-induced chromosome aberrations in human CD34+ cells *Radiat. Res.* **167** 541–50

Barendsen G W 1968 Responses of cultured cells, tumours and normal tissues to radiations of different linear energy transfer *Curr. Topics Radiat. Res.* **4** 293–356

Belli M *et al* 2000 Inactivation of human normal and tumour cells irradiated with low energy protons *Int. J. Radiat. Biol.* **76** 831–9

Carabe A, Moteabbed M, Depauw N, Schuemann J and Paganetti H 2012 Range uncertainty in proton therapy due to variable biological effectiveness *Phys. Med. Biol.* **57** 1159–72

Elsässer T, Krämer M and Scholz M 2008 Accuracy of the local effect model for the prediction of biologic effects of carbon ion beams in vitro and in vivo *Int. J. Radiat. Oncol. Biol. Phys.* **71** 866–72

Friedrich T, Scholz U, Elsässer T, Durante M and Scholz M 2013 Systematic analysis of RBE and related quantities using a database of cell survival experiments with ion beam irradiation *J. Radiat. Res.* **54** 414–514

Furusawa Y *et al* 2000 Inactivation of aerobic and hypoxic cells from three different cell lines by accelerated (3)He-, (12)C- and (20)Ne-ion beams *Radiat. Res.* **154** 485–96 erratum 2012 *Radiat. Res.* **177** 129–31

Grassberger C, Trofimov A, Lomax A and Paganetti H 2011 Variations in linear energy transfer within clinical proton therapy fields and the potential for biological treatment planning *Int. J. Radiat. Oncol. Biol. Phys.* **80** 1559–66

Grün R 2015 Assessment of potential advantages of relevant ions for particle therapy: a model based study *Med. Phys.* **42** 1037–47

Grün R, Friedrich T, Krämer M, Zink K, Durante M and Engenhart-Cabillic R *et al* 2015 Assessment of potential advantages of relevant ions for particle therapy: a model based study *Med. Phys.* **42** 1037–47

Hall E J and Giaccia A 2011 Radiobiology for the Radiologist 7th edn (Lippincott: Philadelphia, PA) ch 6 and 7

Hawkins R B 2003 A microdosimetric-kinetic model for the effect of non-Poisson distribution of lethal lesions on the variation of RBE with LET *Radiat. Res.* **160** 61

Hawkins R B 2009 The relationship between the sensitivity of cells to high-energy photons and the RBE of particle radiation used in radiotherapy *Radiat. Res.* **172** 761–76

Inaniwa T, Suzuki M, Furukawa T, Kase Y, Kanematsu N, Shirai T and Hawkins R B 2013 Effects of dose-delivery time structure on biological effectiveness for therapeutic carbon-ion beams evaluated with microdosimetric kinetic model *Radiat. Res.* **180** 44–59

Jones B 2010 The apparent increase in the β-parameter of the linear quadratic model with increased linear energy transfer during particle irradiation *Brit. J. Radiol.* **83** 433–6

Jones B 2015a Towards achieving the full clinical potential of proton therapy by inclusion of LET and RBE models *Cancers (Basel)* **17** 460–80

Jones B 2015b A simpler energy transfer efficiency model to predict relative biological effect (RBE) for protons and heavier ions *Frontiers Oncol.* **11** 184; erratum 2016 *Frontiers Oncol.* **6** 32

Jones B, Carabe-Fernandez A and Dale R G 2007 The oxygen effect *Radiobiological Modelling in Radiation Oncology* (London: British Institute of Radiology) 138–57

Jones B, Underwood T C, Carabe-Fernandez A and Dale R G 2011 Further analysis of fast neutron relative biological effects and implications for charged particle therapy *Brit. J. Radiol.* **84** 11–8

Kase Y, Kanai T, Matsufuji N, Furusawa Y, Elsässer T and Scholz M 2008 Biophysical calculation of cell survival probabilities using amorphous track structure models for heavy-ion irradiation *Phys. Med. Biol.* **53** 37–59

Katz R, Ackerson B, Homoyooufar M and Sharma S C 1971 Inactivation of cells by heavy ion bombardment *Radiat. Res.* **47** 402–25

Newhauser W D and Zhang R 2015 The physics of proton therapy *Phys. Med. Biol.* **60** R55–209

Sørensen B S, Overgaard J and Bassler N 2011 In vitro RBE-LET dependence for multiple particle types *Acta Oncol.* **50** 757–62

Todd P 1967 Heavy-ion irradiation of cultured human cells *Radiat. Res.* **7** 196–207

Weyrather W K, Ritter S, Scholz M and Kraft G 1999 RBE for carbon track-segment irradiation in cell lines of differing repair capacity *Int. J. Radiat. Biol.* **75** 1357–64

Wilkens J J and Oelfke U 2004 A phenomenological model for the relative biological effectiveness in therapeutic proton beams *Phys. Med. Biol.* **49** 2811–25

IOP Publishing

Chapter 9

Inclusion of the energy-efficiency LET and RBE model in proton therapy

9.1 Introduction

Radiotherapy has evolved empirically, with most technical developments contributing to better clinical outcomes as either a better cure rate, or reduced complication rate, or both. There are some exceptions such as the increment in dose rate for brachytherapy purposes and fast neutron therapy where clinical results were on the whole less than satisfactory.

The fundamental distinction between early and late tissue reactions, and their different fractionation sensitivities and dose ranges are important and they provide limits to prescribed doses. Recent recognition that late vascular damage is responsible for many chronic effects of radiation and that the relative risk of radiation induced coronary artery disease is estimated to be around 7.5% per Gray (Darby *et al* 2013), as well as the established cancer induction risks of at least 5% per Sievert (Hall and Gaccia 2011), leads to a logical requirement for radiation beams that can markedly reduce integral dose and so the amount of tissue traversed. This is achievable with positively charged particle beams such as protons and light ions, due to their reduced path length in tissues caused by the Bragg peak effect.

Technical developments in radiotherapy need careful assessment of their biological effects, since intended gains can occur along with subtle or even major disadvantages. In the case of proton beam therapy (PBT), concerns remain about the lack of published evidence showing clear-cut improvements in outcomes. Many overviews point out the paucity of publications that show clear-cut improvements in terms of tumour control and/or late normal tissue complications, as well as the lack of high level medical evidence using randomised studies. Justification that proton therapy is safe, by using only paediatric data such as medulloblastoma (Yock *et al* 2016) and ependymoma (Peeler *et al* 2016), should not be extrapolated to adult

doi:10.1088/978-0-7503-1338-4ch9

treatments. This is because lower doses, below the threshold for serious complications such as spinal myelitis and brain necrosis, are used in treating this tumour type.

These concerns have received further support from some publications. There are reports of higher than expected radiological changes in ependymoma and meningioma patients after proton therapy (Peeler *et al* 2016, Mohan 2017), as well as markedly increased fibrosis occurring at the distal portions of proton spread-out Bragg peaks, compatible with relative biological effect (RBE) values of 3–4, which recalls the values found with fast neutrons (Underwood and Paganetti 2016). Also, there is some concern within the clinical community regarding results published from Switzerland's prestigious PSI laboratory, with serious neurological toxicity (blindness and cortical brain necrosis) as high as around 12.3% (Weber *et al* 2016), a result that would not be tolerated with photon-based therapy, where protocols are designed not to allow more than 1–2% risks, and often aim for below this range, while being capable of delivering high tumour doses using modern photon techniques. Also, the long term results of children treated with protons in Japan show a significant rate of late complications; the grade 3 or higher late toxicities were 6%, 17% and 17% at five, ten and twenty years, respectively (Mizumoto *et al* 2016). As far as tumour control rates are concerned there is no clear-cut advantage from the use of protons for intracranial tumours, as reviewed by Combs (2017). Further clinical examples are given below.

Another barrier appears to be that Bragg peak placement issues dominate discussions, with minimal attention to the radiobiology uncertainties, so that it becomes difficult to determine the cause of toxicity or failure to cure. Realistic assessment of both are required, but it cannot be disputed that the range of known variation in the RBE far exceeds that of Bragg peak positioning. This aspect must also be understood by the inevitable fact that displacement of the Bragg peak position away from the ICRU target volumes might increase normal tissue toxicity outside the defined planning target volume, but would be expected to be accompanied by reduced tumour control; there is no evidence that this has occurred, so it is reasonable to invoke the RBE as being the main culprit.

The RBE is of paramount importance since it determines the dose given to the patient in equivalent-Gy. Essentially, if the RBE is incorrect the proton dose may be too low or too high, with consequent effects in each patient.

The RBE depends on multiple factors, as discussed in the previous chapter. This multifactorial basis of the RBE implies that it cannot be a constant value, but must be a continuous variable (as is height and weight in humans). Particularly relevant are the radiobiological properties of the control low linear energy transfer (LET) of conventional megavoltage radiotherapy that govern the numerator of the RBE definition. ICRU 16 (1970) states that the mean LET for megavoltage radiations (electrons and photons) are around 0.2–0.3 keV μm^{-1}. From the shapes of the cell survival or bio-effect curves at low and high LET, the numerator dose will increase to a greater extent than the denominator dose for a given iso-effect when the dose per fraction is changed. This can be seen in figure 2.1 in chapter 2, with numerous practical examples in the preceding chapter.

Proton therapy has two potential Achilles heels (unlike only one susceptible heel in the original Homeric poem). They are (1) physical dose placement uncertainty and (2) biological effect (RBE) uncertainties. First it is necessary to describe some of the physical uncertainties before considering the RBE issues in greater detail. The physical issues have already been discussed in chapter 1.

9.2 RBE uncertainties

The RBE concept is so often either misunderstood or underestimated in terms of complexity. It is vital to understand its multifactorial dependency, already listed above. All clinicians should understand that if the RBE is incorrect, so also will the dose given to the patient, sometimes by a percentage change that can exceed normally acceptable treatment plan dose variations, or the legally permitted range of dose due to errors in beam delivery (Dale *et al* 2009). The RBE also has an impact on proton range uncertainties (Carabe *et al* 2012). A reported reduction in RBE at long range (Calugaru *et al* 2011) is possibly connected with the increasingly separating tracks, as discussed in chapter 1, and has important implications that need further experimentation.

Proton RBEs have similarities with those for heavier ions than protons, but with some differences in scale. High proton RBEs have been found in various experiments, in high LET parts of the beam (Belli *et al* 2000, Britten *et al* 2013, Marshall *et al* 2016). Also, the rise in the RBE per unit increase in LET is larger at lower LET values for protons compared to all heavier ions (probably because of the smaller 'Katz radius' of ionisation associated with each track (see chapter 1), which is directly proportional to the nuclear charge (Z), and protons have the lowest Z number of 1. So, protons are more efficient at increasing the RBE at lower values of LET. This causes the turnover of the RBE with LET at around 30.5 in Belli's experiments (see figure 9.1), rather than at between 100–120 and 180–220 keV μm^{-1} for protons, helium and carbon ions, respectively, as shown in chapter 8. The potential for high RBE values occurring at deceptively low LET values consequently arises.

The essential clinical features of RBE are its inevitable dependency on LET, dose and tumour/tissue type. Changes in proton RBE with dose per fraction and tissue type have not been as extensively researched when compared with the range of experimental systems used for fast neutrons. In the latter case, there was a clear dependency on tissue type; fast proliferating, acute reacting tissues (with high α/β ratios) had much smaller changes in RBE with dose than was the case for far more slowly proliferating late reacting tissues (with low α/β ratios) (Carabe-Fernandez *et al* 2007, 2010). The collected proton experiments, and the expedient decision to allocate an overall RBE of 1.1, representing a median value, to all tissues and tumours (regardless of their α/β ratio), is discussed in chapter 7), and in order for proton therapy to improve must be reconsidered. This should be coupled with more detailed experimental studies and analysis of human clinical outcomes, including radiological changes in normal tissues for LET-RBE effects.

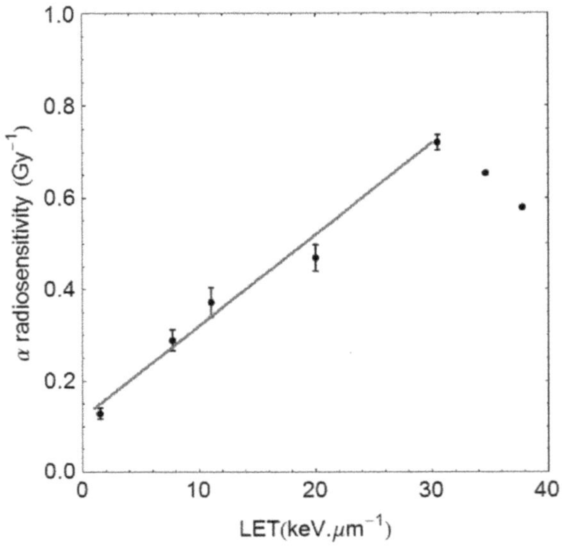

Figure 9.1. Data of Belli *et al* (2000) showing α rising linearly with LET. The least-squares fitted line is α_H = 0.12 + 0.02 LET for data points below 31 keV μm^{-1}, which will only be the case for the strain of V-79 cells used, where 0.12 Gy^{-1} is the control low LET α_L value. Reproduced from Jones (2015a).

9.3 Description of the quantitative model

To determine the RBE the same equations as in chapter 8 are used, but with the proton z number of 1. Further details can be found in two publications (Jones 2015a, Jones 2017) concerning the link between the low LET reference values for α and β and their high LET counterparts, and in the supplementary data in the most recent reference.

The efficiency of cell killing, in terms of the α and β parameters, are both assumed to rise linearly with LET to a maximum at 30.5 keV μm^{-1} and to a level determined by the low LET (control radiation) α and β parameters (respectively termed α_L and β_L, but where α/β is used it will refer to the low LET state). The model considers increments in α and β that are proportional to LET, but using a saturation effect that limits the maximum possible increase in α and β, as given in chapter 8.

Since the proton LET–RBE turnover point (defined as LET_U) is at 30.5 keV μm^{-1} (Belli *et al* 2000), and the increments in RBE (and consequently the radiosensitivity parameters) with LET appear to be linear (Belli *et al* 2000, Sørensen *et al* 2011) then with increasing dose the overall RBE will fall in proportion to the shift between the high LET α and β parameters, as shown in figure 9.2 (a more complete version of this graphic is given as figure 8.1 in chapter 8).

The model outputs, using a baseline (control) LET of 0.22 keV μm^{-1} for megavoltage photons, are now displayed in the form of figures 9.3(a)–(e), where the fixed parameter was β = 0.03 Gy^{-2} (at the reference low LET), and tables 9.1–9.5, where stochastic estimates are obtained by assuming mean α values of 0.07 Gy^{-1}, for α/β of 2 and 3 Gy for late-reacting normal tissues, and reference radiation α values of 0.275, 0.375 and 0.55 Gy^{-1} for tumours with α/β ratios of, respectively,

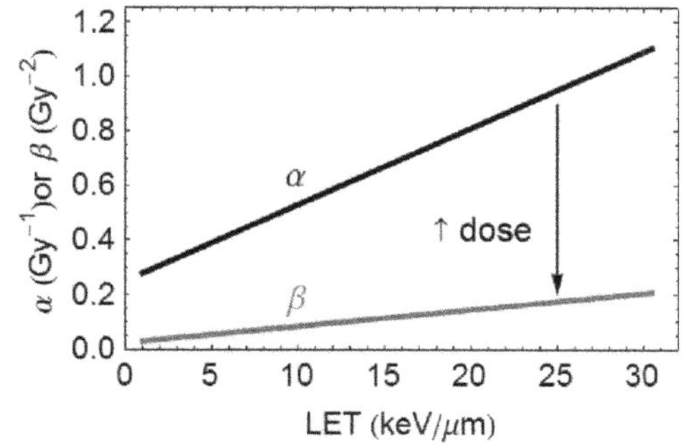

Figure 9.2. Schematic diagram of the linear increase change in α and β with LET. Increasing the dose will increase the proportion of β mediated cell kill and so cause a lower overall RBE (itself determined by the increments in α (changing RBE_{max}) and in β (changing RBE_{min}), and so the relative proportions of RBE_{max} and RBE_{min} that contribute to the overall RBE at any value of LET within this range. The increase in β with LET is always smaller than that in α. So, with increasing dose, the overall RBE follows the direction of the arrow. The numbers used in this figure are not intended to represent any particular cell system. Reproduced from Jones (2015a).

4.5, 10 and 25 Gy. In all cases a standard deviation of 10% of the mean was used in Mathematica random sampling procedures using 1000 samples, which provide the mean RBE and the 1% and 99% quantile ranges.

The steps taken to estimate the RBE are as follows.

1. The RBE calculation is based on the following scaling approach, an extension of the approach taken by Wilkens and Oelfke (2004) as in Jones (2015a and 2017), which includes the relevant parameters.

$$\alpha_H = \alpha_L + \frac{LET_x - LET_C}{LET_U - LET_C}(\alpha_U - \alpha_L)$$

$$\alpha_H = \alpha_L + \frac{LET_x - LET_C}{LET_U - LET_C}\cdot(\alpha_U - \alpha_L)$$

with LET_U, the slope of α and β with LET, is found from the maximum radiosensitivity values:

$\alpha_U = Q/j\ (1-\exp[-j\alpha_L)$, and

$\beta_U = (R/u)(1-\exp[-u\ \beta_L])$.

If LET_U is not known, then use the highest available LET (let this be LET_S and use this to replace LET_U, provided the radiosensitivity values at LET_S are known, which then replace α_U and β_U.

RBE is obtained by solving for d_L in the general iso-effect equation: $\alpha_L\ d_L + \beta_L\ d_L^2 = \alpha_H\ d_H + \beta_H\ d_H^2$, and then dividing by the d_H to give:

$$RBE = \frac{-\alpha_L + \sqrt{\alpha_L^2 + 4\,\alpha_H\beta_L d_H + 4\beta_H\beta_L d_H^2}}{(2\beta_L d_H)}.$$

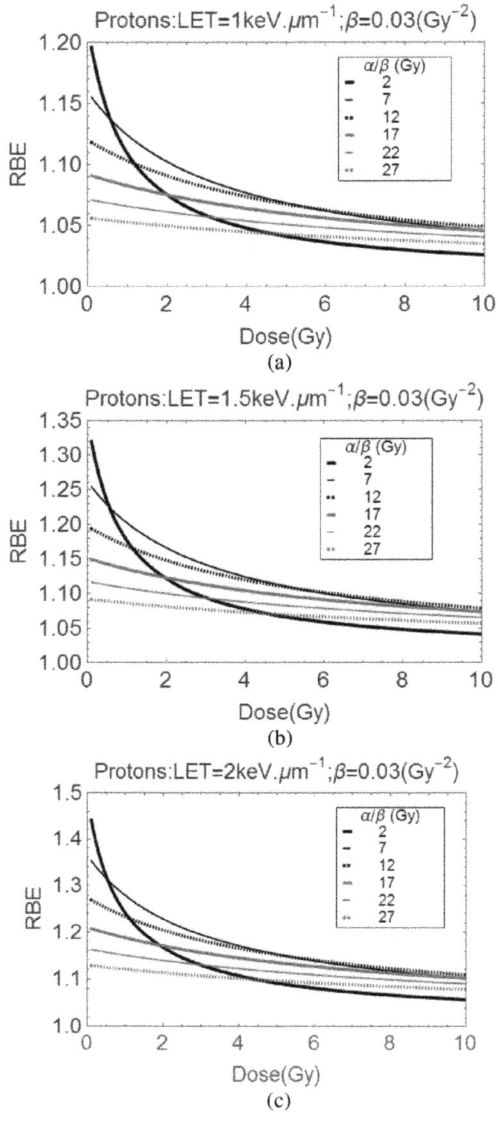

Figure 9.3. (a)–(e) Plots of proton RBE with dose per fraction at specified LET values. Reproduced from Jones (2015a).

The estimates in the tables and figures 9.3 (a)–(e) all show increasing RBE with LET and an inverse relationship between the RBE and dose per fraction. Greater complexity is introduced by varying the α/β ratios. It again can be seen that increases in LET above, say, 1.5 keV μm^{-1}, can be accompanied by significant increments in RBE. The lowest α/β ratio systems give the largest RBE at near zero dose, but then the values fall sharply, the rate of change depending on α/β. High α/β ratio systems have much flatter curves, showing less variation in RBE with dose per fraction. The curves cross over such that larger RBE values can exist for systems

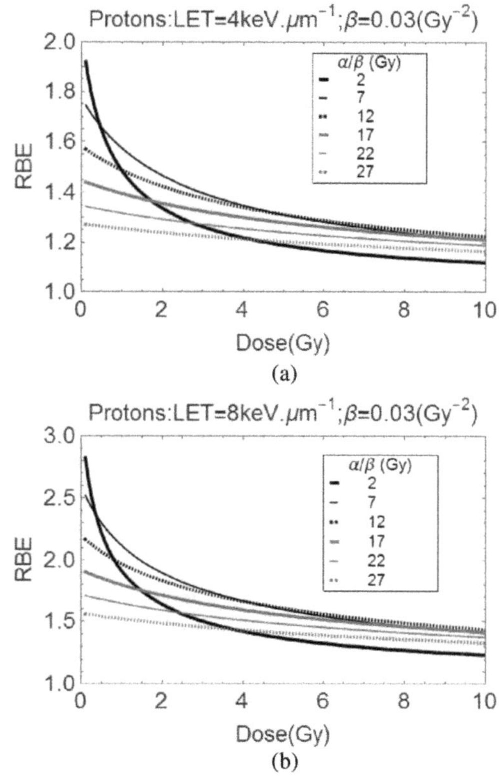

Figure 9.3. (Continued.)

with high α/β ratios compared with low α/β ratios when dose per fraction is sufficiently large; it is important to note that the lowest RBE values then occur in the most fraction-sensitive tissues, suggesting that high dose per fraction can be relatively protective. This is because of enhanced repair capacity and fraction sensitivity in the case of low α/β ratios for the photon (numerator) part of the RBE definition and the converse for the high LET denominator dose. Also, cells with low α/β, with the highest RBE_{max}, will asymptotically approach RBE_{min} with dose at a faster rate owing to the higher proportion of β related cell killing than in cells where α/β is high.

By solving for clinically relevant iso-effects (50 Gy in 25 fractions for spinal cord and optic chiasm, etc, and 60 Gy in 30 fractions for cortical brain in the BED equations with LET–RBE effects included (see chapter 6), the total doses required to maintain each iso-effect are given in figures 9.4(a) and (b), respectively.

Many of the RBE values shown exceed the conventional value of 1.1. Of particular concern is that for low α/β ratio systems such as brain spinal cord and late reactions, the RBE values are highest when the dose per fraction is small. This finding represents a warning to clinicians, who should either consider using a suitably modified RBE, or change the tolerance of key organs at risk in such

Table 9.1. $\alpha/\beta = 2$ Gy: central nervous system late effects with mean $\alpha = 0.07\ Gy^{-1}$ (SD = 0.02). Reproduced from Jones (2017).

Dose (Gy)	LET = 1	LET = 1.25	LET = 1.5	LET = 1.75	LET = 2.0	LET = 4.0	LET = 8.0
$d = 1.25$	1.10(1.08, 1.11)	1.12(1.08, 1.14)	1.15(1.13, 1.18)	1.18(1.16, 1.21)	1.21(1.18, 1.24)	1.42(1.37, 1.48)	1.80(1.7, 1.9)
$d = 1.5$	1.09(1.07, 1.10)	1.11(1.10, 1.13)	1.14(1.12, 1.16)	1.17(1.14, 1.19)	1.19(1.16,1.22)	1.38(1.33,1.44)	1.72(1.63, 1.82)
$d = 1.8$	1.08(1.07, 1.09)	1.10(1.09, 1.12)	1.13(1.11, 1.15)	1.15(1.13, 1.17)	1.17(1.15, 1.20)	1.35(1.30, 1.40)	1.66(1.57, 1.75)
$d = 2$	1.07(1.06, 1.09)	1.10(1.08, 1.11)	1.12(1.10, 1.14)	1.14(1.12,1.16)	1.16(1.14, 1.19)	1.33(1.28, 1.38)	1.62(1.53, 1.71)
$d = 2.5$	1.06(1.05, 1.08)	1.08(1.07, 1.10)	1.10(1.09, 1.12)	1.12(1.10, 1.15)	1.14(1.12, 1.17)	1.29(1.24, 1.34)	1.54(1.46, 1.64)
$d = 3$	1.06(1.05, 1.07)	1.07(1.06, 1.09)	1.09(1.07, 1.11)	1.11(1.09, 1.13)	1.13(1.10, 1.15)	1.25(1.21, 1.31)	1.48(1.41, 1.58)
$d = 5$	1.04(1.03, 1.05)	1.05(1.04, 1.07)	1.06(1.05, 10.8)	1.08(1.06, 1.10)	1.09(1.07, 1.11)	1.18(1.14, 1.23)	1.35(1.28, 1.44)
$d = 10$	1.02(1.01, 1.03)	1.03(1.02, 10.5)	1.04(1.03, 1.06)	1.05(1.03, 1.07)	1.05(1.04, 1.08)	1.11(1.08, 1.12)	1.22(1.15, 1.31)
$d = 12.5$	1.02(1.01, 1.03)	1.03(1.02, 1.04)	1.03(1.02, 1.05)	1.04(1.02, 1.06)	1.05(1.03, 1.07)	1.10(1.06, 1.15)	1.19(1.12, 1.28)

Table 9.2. $\alpha/\beta = 3$ Gy: general normal tissue late effects with mean $\alpha = 0.07\ Gy^{-1}$ (SD = 0.02). Reproduced from Jones (2017).

Dose (Gy)	LET = 1	LET = 1.25	LET = 1.5	LET = 1.75	LET = 2.0	LET = 4.0	LET = 8.0
$d = 1.25$	1.12(1.10, 1.13)	1.15(1.13, 1.17)	1.19(1.17, 1.21)	1.23(1.20, 1.25)	1.26(1.23, 1.29)	1.52(1.46, 1.58)	1.98(1.88, 2.09)
$d = 1.5$	1.11(1.09, 1.12)	1.14(1.12, 1.16)	1.18(1.15, 1.20)	1.21(1.18, 1.23)	1.24(1.21, 1.27)	1.48(1.42, 1.54)	1.91(1.80, 2.01)
$d = 1.8$	1.10(1.09, 1.11)	1.13(1.11, 1.15)	1.16(1.14, 1.18)	1.19(1.17, 1.22)	1.22(1.19, 1.25)	1.44(1.39, 1.50)	1.83(1.73, 1.93)
$d = 2$	1.09(1.08, 1.11)	1.12(1.11, 1.14)	1.15(1.13, 1.17)	1.18(1.16, 1.20)	1.21(1.18, 1.24)	1.42,(1.37, 1.47)	1.79(1.69, 1.88)
$d = 2.5$	1.08(1.07, 1.10)	1.11(1.09, 1.13)	1.14(1.12, 1.16)	1.16(1.14, 1.18)	1.19(1.16, 1.21)	1.37(1.32, 1.43)	1.70(1.61, 1.79)
$d = 3$	1.08(1.06, 1.09)	1.10(1.08, 1.11)	1.12(1.10, 1.14)	1.15(1.12, 1.17)	1.17(1.14, 1.19)	1.34(1.29, 1.39)	1.64(1.55, 1.73)
$d = 5$	1.06(1.05, 1.07)	1.07(1.06, 1.09)	1.09(1.07, 1.10)	1.11(1.09, 1.13)	1.13(1.10, 1.50)	1.25(1.21, 1.30)	1.48(1.40, 1.56)
$d = 10$	1.04(1.03, 1.05)	1.05(1.04, 1.06)	1.06(1.04, 1.07)	1.07(1.05, 1.09)	1.08(1.06, 1.10)	1.17(1.13, 1.21)	1.32(1.25, 1.40)
$d = 12.5$	1.03(1.02, 1.04)	1.04(1.03, 1.05)	1.05(1.04, 1.07)	1.06(1.04, 1.08)	1.07(1.05, 1.09)	1.15(1.11, 1.19)	1.28(1.21, 1.36)

Table 9.3. $\alpha/\beta = 4.5$ Gy: tumours with mean $\alpha = 0.275 \text{Gy}^{-1}$ (SD = 0.0275). Reproduced from Jones (2017).

Dose (Gy)	LET = 1	LET = 1.25	LET = 1.5	LET = 1.75	LET = 2.0	LET = 4.0	LET = 8.0
$d = 1.25$	1.09(1.08, 1.10)	1.12(1.10, 1.13)	1.14(1.12, 1.16)	1.17(1.15, 1.20)	1.20(1.17, 1.23)	1.41(1.35, 1.46)	1.79(1.69, 1.89)
$d = 1.5$	1.08(1.07, 1.09)	1.11(1.09, 1.12)	1.13(1.11, 1.15)	1.16(1.14, 1.18)	1.19(1.16, 1.21)	1.38(1.33, 1.43)	1.73(1.64, 1.82)
$d = 1.8$	1.08(1.07, 1.09)	1.10(1.09, 1.11)	1.12(1.11, 1.14)	1.15(1.13, 1.17)	1.17(1.15, 1.19)	1.35(1.30, 1.40)	1.68(1.59, 1.76)
$d = 2$	1.07(1.06, 1.08)	1.10(1.08, 1.11)	1.12(1.10, 1.14)	1.14(1.12. 1.16)	1.16(1.14, 1.19)	1.33(1.29, 1.38)	1.64(1.56, 1.73)
$d = 2.5$	1.07(1.06, 1.08)	1.09(1.07, 1.11)	1.11(1.09, 1.12)	1.13(1.11, 1.14)	1.15(1.12, 1.17)	1.30(1.26, 1.34)	1.58(1.50, 1.65)
$d = 3$	1.06(1.05, 1.07)	1.08(1.07, 1.09)	1.10(1.08, 1.11)	1.11(1.10, 1.13)	1.13(1.11, 1.15)	1.27(1,23, 1.31)	1.52(1.49, 1.59)
$d = 5$	1.04(1.03, 1.05)	1.06(1.05, 1.07)	1.07(1.06, 1.08)	1.08(1.07, 1.10)	1.10(1.08, 1.11)	1.20(1.17, 1.23)	1.38(1.33, 1.44)
$d = 10$	1.03(1.02, 1.03)	1.03(1.03, 1.04)	1.04(1.03, 1.05)	1.05(1.04, 1.06)	1.06(1.04, 1.07)	1.12(1.09, 1.14)	1.23(1.19, 1.28)
$d = 12.5$	1.02(1.02, 1.03)	1.03(1.02, 1.03)	1.03(1.03, 1.04)	1.04(1.03, 1.05)	1.05(1.04, 1.06)	1.10(1.08, 1.12)	1.19(1.15, 1.23)

Table 9.4. $\alpha/\beta = 10$ Gy: moderately radiosensitive tumours with mean $\alpha = 0.375 \text{ Gy}^{-1}$ (SD = 0.0375). Reproduced from Jones (2017).

Dose (Gy)	LET = 1	LET = 1.25	LET = 1.5	LET = 1.75	LET = 2.0	LET = 4.0	LET = 8.0
$d = 1.25$	1.09(1.8, 1.11)	1.12(1.10, 1.14)	1.15(1.13, 1.18)	1.18(1.15, 1.21)	1.21(1.18, 1.25)	1.44(1.37, 1.51)	1.87(1.73, 2.01)
$d = 1.5$	1.09(1.07, 1.11)	1.12(1.11, 1.14)	1.15(1.12, 1.17)	1.18(1.15, 1.21)	1.20(1.17, 1.24)	1.42(1.35, 1.49)	1.83(1.70, 1.96)
$d = 1.8$	1.09(1.07, 1.10)	1.11(1.09, 1.13)	1.14(1.12, 1.16)	1.17(1.14, 1.20)	1.19(1.16, 1.23)	1.40(1.34, 1.47)	1.79(1.66, 1.92)
$d = 2$	1.08(1.07, 1.10)	1.11(1.09, 1.13)	1.14(1.11, 1.16)	1.16(1.14, 1.19)	1.19(1.16, 1.22)	1.39(1.33, 1.45)	1.76(1.64, 1.89)
$d = 2.5$	1.08(1.07, 1.09)	1.09(1.10, 1.12)	1.13(1.11, 1.15)	1.15(1.13, 1.18)	1.18(1.15, 1.20)	1.36(1.30, 1.42)	1.71(1.60, 1.82)
$d = 3$	1.07(1.06, 1.09)	1.10(1.08, 1.11)	1.12(1.10, 1.14)	1.14(1.12, 1.17)	1.17(1.14, 1.20)	1.34(1.28, 1.40)	1.66(1.55, 1.77)
$d = 5$	1.06(1.05, 1.07)	1.08(1.06, 1.09)	1.10(1.08, 1.11)	1.12(1.10, 1.14)	1.13(1.11, 1.16)	1.28(1.23, 1.32)	1.54(1.45, 1.62)
$d = 10$	1.04(1.03, 1.05)	1.05(1.04, 1.07)	1.07(1.05, 1.08)	1.08(1.06, 1.10)	1.09(1.08, 1.11)	1.19(1.16, 1.23)	1.37(1.30, 1.44)
$d = 12.5$	1.04(1.03, 1.04)	1.05(1.04, 1.06)	1.06(1.05, 1.07)	1.07(1.06, 1.08)	1.08(1.06, 1.10)	1.17(1.14, 1.20)	1.33(1.27, 1.39)

Table 9.5. $\alpha/\beta = 25$ Gy: very radiosensitive tumours with mean $\alpha = 0.55$ Gy^{-1} (SD = 0.055). Reproduced from Jones (2017).

Dose (Gy)	LET = 1	LET = 1.25	LET = 1.5	LET = 1.75	LET = 2.0	LET = 4.0	LET = 8.0
$d = 1.25$	1.08(1.06, 1.10)	1.10(1.08, 1.13)	1.13(1.10, 1.16)	1.15(1.12, 1.19)	1.18(1.14, 1.22)	1.38(1.30, 1.47)	1.77(1.60-1.94)
$d = 1.5$	1.08(1.06, 1.10)	1.10(1.08, 1.13)	1.13(1.10, 1.16)	1.15(1.12, 1.19)	1.18(1.14, 1.22)	1.37(1.29, 1.46)	1.75(1.59, 1.92)
$d = 1.8$	1.08(1.06, 1.09)	1.10(1.08, 1.12)	1.13(1.10, 1.15)	1.15(1.12, 1.18)	1.17(1.14, 1.21)	1.37(1.29, 1.45)	1.73(1.58, 1.90)
$d = 2$	1.08(1.06, 1.09)	1.10(1.08, 1.12)	1.12(1.10, 1.15)	1.15(1.11, 1.18)	1.17(1.13, 1.21)	1.36(1.28, 1.44)	1.72(1.57, 1.88)
$d = 2.5$	1.07(1.06, 1.09)	1.10(1.08, 1.12)	1.12(1.09, 1.15)	1.14(1.11, 1.18)	1.17(1.13, 1.20)	1.35(1.27, 1.43)	1.69(1.55, 1.85)
$d = 3$	1.07(1.06, 1.09)	1.09(1.07, 1.12)	1.12(1.09, 1.14)	1.14(1.11, 1.17)	1.16(1.13, 1.20)	1.34(1.26, 1.41)	1.67(1.53, 1.82)
$d = 5$	1.06(1.05, 1.08)	1.08(1.07, 1.10)	1.10(1.08, 1.13)	1.12(1.10, 1.15)	1.14(1.11, 1.18)	1.30(1.24, 1.37)	1.59(1.47, 1.72)
$d = 10$	1.05(1.04, 1.06)	1.07(1.05, 1.08)	1.08(1.07, 1.10)	1.10(1.08, 1.12)	1.12(1.09, 1.14)	1.24(1.19, 1.30)	1.47(1.38, 1.58)
$d = 12.5$	1.05(1.04, 1.06)	1.06(1.05, 1.08)	1.08(1.06, 1.10)	1.09(1.07, 1.11)	1.11(1.08, 1.13)	1.22(1.17, 1.27)	1.44(1.35, 1.53)

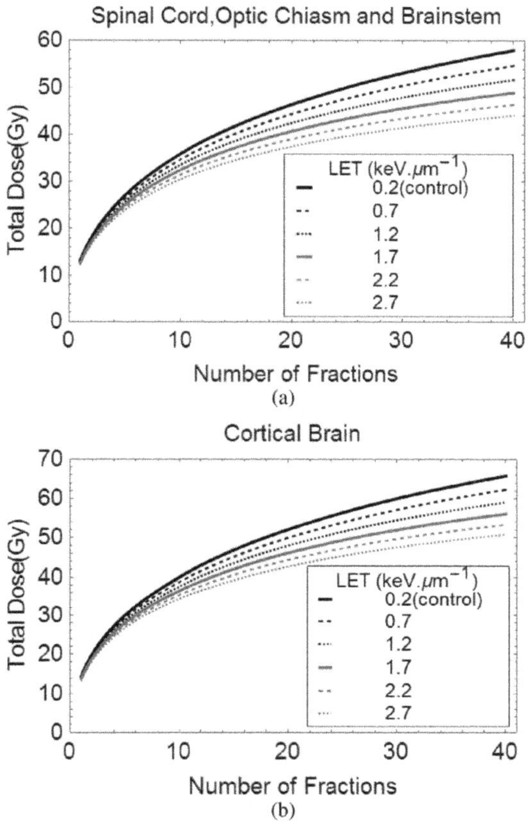

Figure 9.4. Estimated total dose and number of fractions iso-effective with (a) 50 Gy in 25 fractions, and (b) 60 Gy in 30 fractions for the control megavoltage photon (x-ray) treatment with LET of 0.2 keV μm^{-1}. The reduction of total dose needed to maintain these iso-effects is plotted for different LET values commonly encountered in proton therapy.

situations. Recognition of changes in RBE with tissue and tumour α/β ratios, and appropriate use of these ratios to the clinical setting will be important in PBT.

The method of estimation has assumed firm knowledge of the LET_U value for protons. The sensitivity of this value is further explored in chapter 11. Should the slope of the radiosensitivities, or RBE, with increasing LET be known by experiment or clinical data, it would be possible to estimate the RBE more directly, as mentioned in chapter 8 (section 8.3.5). Without such data, the more generic method has to be pursued or alternative approaches based on the low LET α/β ratio, although such a method may incur some inaccuracies for high dose per fraction if β is assumed to be invariant with LET.

9.4 Some comparisons with experimental data sets

The data by Britten *et al* (2013) was used to compare experimentally determined RBE values with model predictions, with results shown in table 9.6 (where only the

Table 9.6. Hep2 cells (Britten *et al* 2013). The first four comparisons are for the 87 MeV incident energy and the final three for the 200 MeV data. The prediction is accurate to around ±3.5%.

LET (keV μm^{-1})	5.3	9	20.5	28.8	7.8	11	13.6
Predicted RBE*	1.17	1.32	1.81	2.18	1.27	1.40	1.51
Actual RBE**	1.22	1.31	1.75	1.92	1.42	1.55	1.63

*Using slope data and not the α_U concept.
**RBE values in publication are divided by 1.2 to avoid potential inaccurate values due to conversion to cobalt-equivalent RBE, and an assumed control LET of 1 keV μm^{-1} was used (120 keV x-rays).

Table 9.7. RBE estimations of Hep2 cells (Britten *et al* 2013). The first four comparisons are for the 87 MeV incident energy and the final three for the 200 MeV data. These predictions are accurate to around 22%, but reduces to 7.5% for LET values below 14 keV μm^{-1}.

LET (keV μm^{-1})	5.3	9	20.5	28.8	7.8	11	13.6
Predicted RBE*	1.27	1.54	2.46	3.16	1.45	1.69	1.89
Actual RBE**	1.22	1.31	1.75	1.92	1.42	1.55	1.63

*Using the α_U concept and assumed LET_U at 30.5 keV μm^{-1}.
**RBE values in publication are divided by 1.2 to avoid potential inaccurate values due to conversion to cobalt-equivalent RBE, and an assumed control LET of 1 keV μm^{-1} was used (120 keV x-rays).

slope of the radiosensitivities are used, with β being extremely small) and table 9.7 (where the full LET_U and α_U assumptions are used). Because the RBE values were published as cobalt equivalent RBE at 0.1 survival fraction using a simple 1.2 conversion from the 120 KeV x-ray control radiation data, the unmodified data was used since the 1.2 could be erroneous. The problem then arises as to what was the most relevant value of LET in the control beam, there being no mention of filtration. The estimated RBE will depend on this value, and 1 keV μm^{-1} has been assumed, which provides an accuracy of 22% due to systematic over-estimation, but this reduces to 8.4% for LET values below 14 keV μm^{-1}. This raises the question as to the accuracy of the LET_U and α_U value used. Pragmatically, the α_U value could be reduced by the degree of inaccuracy to compensate for the systematic shift. If the LET_U value arrived at from simplistic consideration of the Belli *et al* data is incorrect, and if higher turnover position values are used, the error is reduced, as shown in figure 9.5, where the control irradiation LET is also varied. This approach would suggest that the LET_U may be between 45 and 55 keV μm^{-1}. Then, for example, the error falls to around 5% for a control LET of 1.25 and $LET_U = 45$ keV μm^{-1} Further comprehensive experiments are indicated to determine a definitive value and are discussed further in chapter 11.

Such data, as well as many other examples in the literature, are hampered by the relatively few data points, imprecision about the actual position of LET_U and lack of standardisation to megavoltage radiation for control experiments. Further discussion of this problem and how research may better identify LET_U is given in the final chapter.

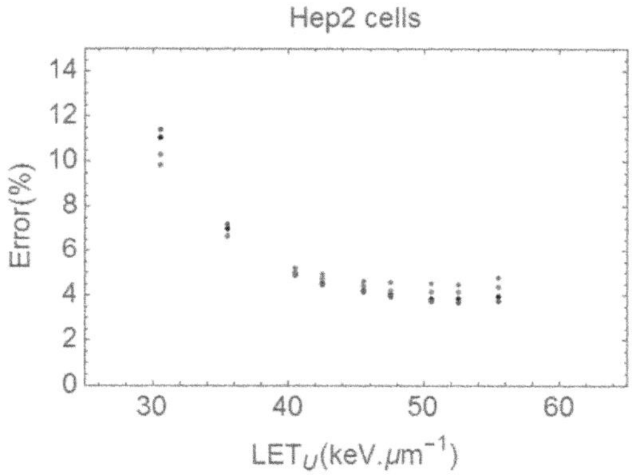

Figure 9.5. Estimation of error in predicting RBE with assumed value of LET_U, with further variations in assumed control LET (coded as blue, black, red and grey for 0.75, 1, 1.5 and 2 keV μm^{-1}, respectively).

9.5 Two clinical examples where PBT could be suboptimal

In addition to the clinical example already given in figure 9.2, the following two clinical entities are relevant.

9.5.1 Prostate cancer

Most American proton centres have based their business cases on treating high numbers of prostate cancer patients. There are some concerns about reports of enhanced side effects, but the outcomes are difficult to establish from small numbers in uncontrolled studies (Jones 2015a). The following list of techniques and assumptions could contribute to enhanced toxicity and have been used in some/all treatment centres.

- To reduce gantry seep and repositioning time, one field per day treatments and use of only two field plans, both inevitably increase the dose per fraction outside the high dose treatment volume.
- Reduced beam shaping and conformity indices, especially with passive scattering, when compared to best photon techniques; this may be improved by scanned beams at the expense of closer inter-track distances (see chapter 1).
- RBE values for late complications will probably exceed 1.1 in spread-out Bragg peaks, so the biological doses to relevant volumes of small bowel and rectum will exceed those with photons.

It remains to be determined if better results emerge with scanned beams from all portals per day, with an emphasis on maximising LET in the GTV, and using a higher RBE allocation of, say, 1.2 for normal tissues.

9.5.2 Paediatric cancers and other radiosensitive tumours such as lymphomas

Here, the tumour α radiosensitivity parameter is as high, resulting in α/β of 28 Gy (see chapter 2), leading to low RBEs, sometimes less than the 1.1 used for protons. In such cases, where RBEs may be as low as, for example, 1.03–1.07, the use of 1.05 or no RBE at all would, respectively, reduce or eliminate the chance of under-dosage. Such an approach would require careful assessment of the normal tissues close to the tumours, since they might then be overdosed, but in many cases of radiosensitive tumours this should be acceptable, since the relatively low curative doses are within normal tissue tolerance for severe late effects. Random sampling simulations of tumour control probabilities, including numbers of patients required to reach significance, are given elsewhere (Jones 2014).

9.6 Prediction of tumour response from the RBE increment

It is always tempting to suggest that, at low dose per fraction, the reference low-LET α/β ratio will closely indicate the likely RBE, since this assumption is built into many proton RBE models. It is pertinent to mention some fast neutron data (Warenius and Britten, 1994, Warenius *et al* 1994) in 30 human cell lines; the rank order of surviving fraction data after 2 Gy (photon) and 1.6 Gy (neutron) exposures, respectively the SF_2, and $SF_{1.6}$, were not the same for photons and neutrons, but a trend was noticed in that the most resistant cells seem to gain most in RBE. Since neutrons mainly ionise by forming recoil protons these results are probably relevant to proton treatments.

A simulation of proton RBE for a typical range of human cancers is given in figure 9.6. Here two random samples are generated (mean $\alpha = 0.275$ Gy^{-1}, SD = 0.07; mean $\beta = 0.03$, SD = 0.008), and both parameters were ordered in lists from lowest to highest. In figure 9.6(a), the RBEs are then estimated for each pair of parameters, using the model used above with separate increments in α and β with LET. It can be seen that, although showing a general trend with RBE, the rank order of α/β will not determine the RBE ranking. This is because of the different magnitudes of the increments in α and β with LET, although being greatest for the most radioresistant cells, but with saturation of this effect for the most radiosensitive cells, and shows the importance of using both α and β independently, since the use of α/β as the primary input to determine RBE, shown in figure 9.6(b), perfectly maintains the rank order from low to high α/β with the corresponding RBE values, which is inconsistent with the data quoted above. Further comparisons of the different proton models available is being prepared by Rorvik in Norway, who is finding considerable variation in the use of biological data and the estimated RBE values depending on the assumptions made in the model.

9.7 Concluding discussion

Although proton therapy is the most promising external beam therapy method, it has its limitations. The obvious advantages of reduced or absent radiation exposure over wide tissue volumes are that many acute and late side effects are inevitably reduced. In the opinion of the present author, the physical and radiobiological uncertainties in tissues within and close to the target volumes could possibly be sufficient to cause unexpected late effects in some patients, by changing the delivered biological effective dose.

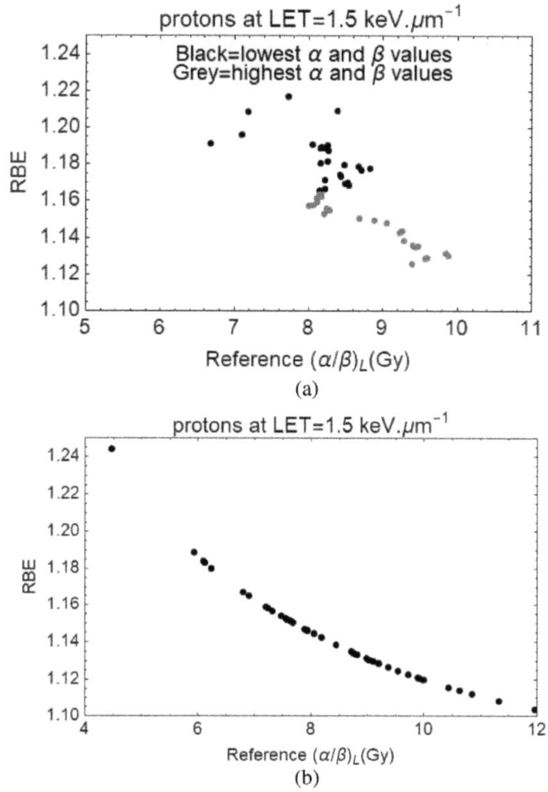

Figure 9.6. Plot of estimated RBE against the low LET reference radiation α/β using the assumptions given in the text (a) in the case where α and β are LET-modified independently and (b) where α/β is LET modified, for the LET value shown, which is typical of the mid-spread-out Bragg peak.

The above suggestions, along with more clinical research publications and molecular targeted approaches, which might capitalise on the apparent reduction of RBE due to reduced repair capacity in some tumour types, as well as the better identification of RBE values in late reacting tissues, might further improve proton therapy results. There appears to be wider support for changing the RBE allocations within the proton therapy prescription process (Underwood and Paganetti 2016) and using LET maps (Grassberger *et al* 2011). Suggestions for the use of the product of LET and dose have been criticised because dose is inversely related to RBE, while LET and RBE will in most instances be linearly related; there is also a wide variation in the product when plotted against surviving fraction (Jones 2017). The present writer is of the opinion that the treatment planning process should include RBE effects, estimated from the LET and intended dose, and where important normal tissue structures exist along the distal ends of proton beams then dose reduction should be applied if the RBE is found to indicate a potentially harmful dose (or biological effective dose).

Proton therapy clinicians and physicists must understand the full implications of a switch between photon and proton therapy. It is clear that the large capital investments in PBT equipment should be accompanied by further basic and applied research, with broader education of the multidisciplinary staff involved in its delivery.

References

Belli M *et al* 2000 Inactivation of human normal and tumour cells irradiated with low energy protons *Int. J. Radiat. Biol.* **76** 831–9

Britten R A *et al* 2013 Variations in the RBE for cell killing along the depth-dose profile of a modulated proton therapy beam *Radiat. Res.* **179** 21–8

Calugaru V, Nauraye C, Noël G, Giocanti N, Favaudon V and Mégnin-Chanet F 2011 Radiobiological characterization of two therapeutic proton beams with different initial energy spectra used at the Institut Curie Proton Therapy Center in Orsay *Int. J. Radiat. Oncol. Biol. Phys.* **81** 1136–43

Carabe A, Moteabbed M, Depauw N, Schuemann J and Paganetti H 2012 Range uncertainty in proton therapy due to variable biological effectiveness *Phys. Med. Biol.* **57** 1159–72

Carabe-Fernandez A, Dale R G, Hopewell J W, Jones B and Paganetti H 2010 Fractionation effects in particle radiotherapy: implications for hypo-fractionation regimes *Phys. Med. Biol.* **55** 5685–700

Carabe-Fernandez A, Dale R G and Jones B 2007 The incorporation of the concept of minimum RBE (RBE_{min}) into the linear-quadratic model and the potential for improved radiobiological analysis of high-LET treatments *Int. J. Radiat. Biol.* **83** 27–39

Combs S E 2017 Does proton therapy have a future in CNS tumors? *Curr. Treat. Options Neurol* **19** 12

Dale R G, Jones B and Carabe-Fernandez A 2009 Why more needs to be known about RBE effects in modern radiotherapy *J. Appl. Radiat. Isotopes* **67** 387–92

Darby S C *et al* 2013 Risk of ischemic heart disease in women after radiotherapy for breast cancer *N. Engl. J. Med.* **368** 987–98

Grassberger C, Trofimov A, Lomax A and Paganetti H 2011 Variations in linear energy transfer within clinical proton therapy fields and the potential for biological treatment planning *Int. J. Radiat. Oncol. Biol. Phys.* **80** 1599–66

Grosshans D R 2016 Proton therapy for Medulloblastoma *Lancet Oncol.* **17** 258–259

Hall E J and Giaccia A 2011 *Radiobiology for the Radiologist* 7th edn (Philadelphia, PA: Lippincott) chs 6 and 7

ICRU Report 16 1970 *Linear Energy Transfer* (Bethesda, MD: ICRU)

Jones B 2014 Editorial: patterns of failure after proton therapy in medulloblastoma *Int. J. Radiat. Oncol. Biol. Phys.* **90** 25–6

Jones B 2015a Towards achieving the full clinical potential of proton therapy by inclusion of LET and RBE models *Cancers (Basel)* **7** 460–80

Jones B 2015b A simpler energy transfer efficiency model to predict relative biological effect (RBE) for protons and heavier ions *Front. Oncol.* **5** 184; erratum *Front. Oncol.* **6** 32

Jones B 2017 Clinical radiobiology of proton therapy: modeling of RBE *Acta Oncol* **56** 1374–8

Marshall T I, Chaudhary P, Michaelidesová A, Vachelová J, Davídková M, Vondráček V, Schettino G and Prise K M 2016 Investigating the implications of a variable RBE on proton dose fractionation across a clinical pencil beam scanned spread-out Bragg peak *Int. J. Radiat. Oncol. Biol. Phys.* **95** 70–7

Mizumoto M *et al* 2016 Long-term follow-up after proton beam therapy for pediatric tumors: a Japanese national survey *Cancer Sci.* **108** 444–7

Mohan R 2017 Private communication

Peeler C R, Mirkovic D, Titt U, Blanchard P, Gunther J R, Mahajan A, Mohan R and Grosshans D R 2016 Clinical evidence of variable proton biological effectiveness in pediatric patients treated for ependymoma *Radiother. Oncol.* **121** 395–401

Sørensen B S, Overgaard J and Bassler N 2011 *In vitro* RBE-LET dependence for multiple particle types *Acta Oncol.* **50** 757–62

Warenius H M, Britten R A, Browning P G, Morton I E and Peacock J H 1994 Identification of human *in vitro* cell lines with greater intrinsic cellular radiosensitivity to 62.5 MeV (p-->Be+) neutrons than 4 MeV photons *Int. J. Radiat. Oncol. Biol. Phys.* **28** 913–20

Underwood T and Paganetti H 2016 Variable proton relative biological effectiveness: how do we move forward? *Int. J. Radiat. Oncol. Biol. Phys.* **95** 56–8

Warenius H M and Britten R A 1994 *In vitro* studies of intrinsic cellular radiosensitivity following 4 MeV photons or 62.5 MeV (p-->Be+) neutrons. Potential implications for high LET therapy *Acta Oncol.* **33** 241–9

Weber D C, Malyapa R, Albertini F, Bolsi A, Kliebsch U, Walser M, Pica A, Combescure C, Lomax A J and Schneider R 2016 Long term outcomes of patients with skull-base low grade chondrosarcoma and chordoma patients treated with pencil beam scanning proton therapy *Radiother. Oncol.* **120** 169–74

Wilkens J J and Oelfke U 2004 A phenomenological model for the relative biological effectiveness in therapeutic proton beams *Phys. Med. Biol.* **49** 2811–25

Yock T I *et al* 2016 Long term toxic effects of proton radiotherapy for paediatric medulloblastoma: a phase 2 single arm study *Lancet Oncol.* **17** 287–98

Chapter 10

Compensating for elapsed time: unintended treatment interruptions and re-treatments

10.1 Introduction

This chapter considers two separate time related issues, and the estimation of the dose required to compensate for:

1. continuing tumour cell repopulation during unintended treatment interruptions resulting in extension of the overall treatment time, usually by a few days or weeks,
2. recovery of normal tissue radiation tolerances at extended times between six months to many years after a treatment course, and which may allow further radiation therapy to be given.

10.2 Unintended treatment interruptions

10.2.1 Background

A reduction in tumour control with extended overall treatment times have been shown in classical experimental studies such as Fowler *et al* (1974). The clinical problem of compensating for treatment interruptions were initially modelled using power law equations, but these proved unreliable, since the parameters used were derived from acute reacting tissue studies and inappropriately applied to late reacting tissues; as a result enhanced late toxicities occurred. A more realistic identification of the clinical problem was made by Withers *et al* (1988). Subsequent publications showed a reduction in tumour control by about 1–2% per extended treatment day in the case of squamous cell cancers and transitional cell cancers. Rapid repopulation is now a better understood cause of treatment failure (Kim and Tannock 2005). References for the above facts can be obtained in the report of the UK Royal College of Radiologists (2008). It took a longer time to identify lower control rates with treatment protraction in other forms of cancer, especially those associated with slower tumour cell kinetics such prostate cancer (Thames *et al* 2010),

doi:10.1088/978-0-7503-1338-4ch10

which appear to manifest as around 0.25 Gy day^{-1} after 6–7 weeks. Further clinical data set analysis is likely to produce new information about other classes of tumour in the future.

Modelling is used in these circumstances as the best guide to compensatory dose estimation, since clinical trials that allow deliberate treatment interruptions would not be ethical. The methods use the equations given in chapter 2 (equations (2.6)–(2.10). Especially important is the basic biological effective does (BED) equation, which includes repopulation, but modified for high LET radiations as

$$\mathrm{BED} = nd\left(\mathrm{RBE}_{\max} + \mathrm{RBE}_{\min}^{2}\frac{d}{\left(\frac{\alpha}{\beta}\right)_{\mathrm{L}}}\right) - K_{\mathrm{L}}(T - T_{\mathrm{K}}), \tag{10.1}$$

where T is the overall time, K_L the low linear energy transfer (LET) daily BED repopulation equivalent and T_K a lag time until repopulation is active. This BED equation has been obtained from the basic equation given as equation (2.7) in chapter 2 by dividing throughout by α_L, so the BED units of K_L are the same as for any photon treatment, just as the α/β ratio used, with the relative biological effect (RBE) limits, is that of the low LET condition. This last equation, with these specific parameters, will be the method used below. However, if $(\alpha/\beta)_H$ is used to compute BED without the necessity for using RBE values, then it is essential to use K_H and not K_L values. The alternative repopulation factor will be obtained by dividing by α_H, rather than α_L, so that:

$$K_{\mathrm{H}} = \frac{0.693}{\alpha_{\mathrm{H}} T_{\mathrm{eff}}}, \tag{10.2}$$

and the relationship between K_L and K_H is given by:

$$K_{\mathrm{H}} = \frac{K_{\mathrm{L}}}{\mathrm{RBE}_{\max}} \tag{10.3}$$

and then the BED will be

$$\mathrm{BED} = nd\left(1 + \frac{d}{\left(\frac{\alpha}{\beta}\right)_{\mathrm{H}}}\right) - K_{\mathrm{H}}(T - T_{\mathrm{K}}). \tag{10.4}$$

10.2.2 Treatment delays

The causes of treatment delays are numerous and include accelerator breakdown, patient illness, severe tissue reactions, chemotherapy complications, transport delays, etc. Particle therapy is especially vulnerable, since treatments normally depend on only one accelerator (rather than one per room for x-ray based treatments). For long accelerator or gantry breakdowns, close liaison between the physicist and oncologist is required to discuss alternative treatment approaches such as x-ray IMRT or other focal x-ray based techniques.

10.2.3 Calculations for compensation of treatment interruptions

The calculations are relatively simple but involve many steps and potential pitfalls (Dale *et al* 2002, Bese *et al* 2007). It must be appreciated that exact compensations for either a normal or a tumour tissue iso-effect will lead to a detriment in the other, so it is necessary to provide a range of solutions that can be chosen by the clinician in charge, and—where possible—the views of the patient.

Interruptions occurring early in the treatment course are easier to compensate, for example by working on weekends or by using two well-spaced treatments on some days close to the end of the week. The difficult cases are those interruptions that occur later in the treatment course. A model calculation for a delay in treatment when only part of the treatment was carbon ions has already been published (Jones *et al* 2006), and more briefly given again below. In the case of hadrontherapy, great care is required, especially regarding the units of dose; if calculations involve the true physical dose (for example by conversion from the cobalt equivalent, or equivalent dose, or RBE-Gy) then the answer is in physical dose and any correction would need to be further adjusted to the original units such as RBE-Gy.

The following two worked examples are based on (1) pure proton and (2) pure carbon treatments. Protons are used to show how subtle differences in results are found from the two different calculation methods. The values used are not intended to be specific for any particular tumour and the modelling examples are intended to demonstrate general principles. LET values have not been used in order to reduce the complexity, but the models described in chapters 8 and 9 can be used to translate the working LET values to RBE_{max} and RBE_{min} values, which then allows the RBE to vary according to the specific dose per fraction under consideration. The results are compared with the use of a fixed (constant) RBE.

Worked example 1

A proton treatment of 46 Gy-eq in 23 fractions is prescribed for a paediatric sarcoma affecting the posterior orbit, treating daily five times per week over 31 days; the child does not require anaesthesia only mild sedation. The treatment is interrupted in the fourth week, on the day after 15 fractions had been given and for a duration of seven days.

Assuming that a simple 1.1 RBE factor has been applied to provide the Gy-eq dose, and that the tumour shows constant low rate of repopulation equivalent to 0.3 Gy-eq day^{-1}. Then we use standard BED equations since it is assumed that all doses are convertible by the 1.1 RBE factor. This assumption is used initially and will be compared with a more flexible RBE value using RBE_{max} and RBE_{min} later. For the sake of simplicity, equal doses to critical normal tissue and tumour are assumed; readers can calculate different BEDs for different iso-dose values using the following formula:

$$BED = xnd\left(1 + \frac{xd}{\left(\frac{\alpha}{\beta}\right)}\right) \tag{10.5}$$

where x refers to the change in dose, n is the number of fractions and would be 0.6 for a 60% iso-dose line and 0.35 for a 35% iso-dose line, etc.

10.2.4 Calculations using a fixed RBE value for two different tissue types within PTV

Tissue A: intended BED to planning target volume (PTV) containing normal tissue (NT) late reacting soft tissue using $\alpha/\beta = 3$ Gy, BED is 46 (1 + 2/3) = 76.67 $Gy_{[3]}$.

Tissue B: intended BED to PTV containing NT late reacting brain/neural tissue using $\alpha/\beta = 2$ Gy, BED is 46 (1 + 2/2) = 92 $Gy_{[2]}$.

Tumour: intended tumour BED, using $\alpha/\beta = 10$ Gy is 46(1 + 2/10) − 0.3 T = 45.90 $Gy_{[10]}$.

BEDs before interruptions:

tissue A: 30 (1 + 2/3) = 50 $Gy_{[3]}$
tissue B: 30 (1 + 2/2) = 60 $Gy_{[2]}$
tumour: 30 (1 + 2/10) = 30.3 $Gy_{[10]}$.

The BED deficit is then:

tissue A: 26.67 $Gy_{[3]}$
tissue B: 32 $Gy_{[2]}$
tumour: 15.6 $Gy_{[10]}$.

If uncorrected the tumour BED (over 31 + 7 days) would be

$$= 46(1 + 2/10) - 0.3 \times 38 = 43.8\ Gy_{[10]}.$$

By treating with the same dose per fraction through one weekend, two days can be saved, so the tumour BED is then

$$= 46(1 + 2/10) - 0.3 \times 36 = 44.4\ Gy_{[10]}.$$

By treating twice a day for all remaining weekdays treatments (i.e. eight fractions in four days) the overall time is 32 days and the BED is then

$$= 46(1 + 2/10) - 0.3 \times 32 = 45.6\ Gy_{[10]}.$$

This seems a good alternative and almost completely restores the BED, but may not be feasible if twice-daily sedation is unacceptable.

Another way forward would be to give additional daily fractions, extending overall time further. If, for example, we try to give the remaining compensatory dose in one additional fraction of 2 Gy in one extra day after completing the prescribed dose, with all delays we obtain an overall treatment time of around 39 days.

For $d = 2$ Gy:

$$\text{BED} = 46(1 + 2/10) + d(1 + d/10) - 0.3 \times 39 = 45.9 \text{ Gy},$$

which is almost the same tumour BED as the original prescription, but the NT BEDs will then be tissue A BED = 80 $\text{Gy}_{[3]}$ and tissue B BED = 96 $\text{Gy}_{[2]}$. These have increased. This additional NT BED can be reduced by fractionating the final day of additional treatment into two treatments of 1.1 Gy, so that

$$\text{BED} = 46(1 + 2/10) + 2 \times 1.1(1 + 1.1/10) - 0.3 \times 39.$$

The tumour BED is then 45.94 $\text{Gy}_{[10]}$; tissue A = 79.67 $\text{Gy}_{[3]}$ and tissue B = 95.41 $\text{Gy}_{[2]}$. The NT BEDs have reduced a little.

Consider what might happen if treatment is hypofractionated and given in only three remaining fractions in order to complete in the same overall treatment time but matched to give the same CNS effect in tissue B (a deficit BED of 32 $\text{Gy}_{[2]}$). First, solve $32 = 3 \times d(1 + d/2)$; $d = 3.73$ Gy.

Then, the tumour BED would be lower:

$$\text{BED} = 30(1 + 2/10) + 3 \times 3.73(1 + 3.73/10) - 31 \times 0.3 = 42.06 \text{ Gy}_{[10]}.$$

This is probably not acceptable, so the alternative method given above would be preferred to maintain reasonable tumour control.

Another approach would be to give twice daily treatment for the entire remainder of treatment. In this way, 16 fractions could be given in eight further days after the unintended gap, competing treatment in 36 days. For a dose per fraction of 1.15 Gy, we obtain:
the tumour BED would then be:

$$15 \times 2(1 + 2/10) + 1.15 \times 16(1 + 1.15/10) - 36 \times 0.3 = 45.72 \text{ Gy}_{[10]}$$

the tissue A BED is then:

$$15 \times 2(1 + 2/3) + 1.15 \times 16(1 + 1.15/3) = 75.45 \text{ Gy}_{[3]}$$

and the tissue B BED is:

$$15 \times 2(1 + 2/2) + 1.15 \times 16(1 + 1.15/2) = 88.98 \text{ Gy}_{[2]}.$$

These BEDs all are slightly below the intended BEDs. It should be noted that these twice-daily fractions have not taken into account the possibility of incomplete repair between fractions.

Since the above calculations have been done using a fixed RBE, it should be noted that dose units for the particle therapy are in 'equivalent Gy' rather than Gy.

10.2.5 Calculations using a variable RBE value

In this case the actual physical dose of protons must be used along with the RBE_{max} and RBE_{min}. For the sake of simplicity the repopulation equivalent remains 0.3 Gy day^{-1}, as a physical dose.

The 2 Gy fraction was actually calculated from a 1.1 RBE so the actual dose given is 2/1.1 = 1.82 Gy. If we assume RBE_{max} = 1.15; RBE_{min} = 1.05 for tumour, and a higher value of RBE_{max} = 1.2 for normal tissue and RBE_{max} = 1.3 for brain tissue, but all have the same RBE_{min} (remember that RBE_{min} is mainly operative at high dose per fraction and will have little influence on the calculations below), and we obtain:

tissue A: intended PTV containing normal soft tissue (late-reacting) with $\alpha/\beta = 3$ Gy, the BED is then $= 23 \times 1.82(1.2 + 1.05^2 \times 1.82/3) = 78.23\ Gy_{[3]}$

tissue B: intended PTV containing late reacting brain/neural tissue with $\alpha/\beta = 2$Gy, the BED is then $= 23 \times 1.82(1.3 + 1.05^2 \times 1.82/2) = 96.42\ Gy_{[2]}$

tumour: intended tumour BED, using $\alpha/\beta = 10$ Gy is $23 \times 1.82(1.15 + 1.05^2 \times 1.82/2) - 0.3 \times 31 = 47.24\ Gy_{[10]}$.

The BEDs given before interruptions are:

tissue A: $15 \times 1.82\ (1.2 + 1.05^2 \times 1.82/3) = 51.02\ Gy_{[3]}$
tissue B: $15 \times 1.82\ (1.3 + 1.05^2 \times 1.82/2) = 62.88\ Gy_{[2]}$
tumour: $15 \times 1.82\ (1.15 + 1.05^2 \times 1.82/10) - 0.3 \times 19 = 31.17\ Gy_{[10]}$.

The BED deficit is then:

tissue A: 27.21 $Gy_{[3]}$
tissue B: 33.54 $Gy_{[2]}$
tumour: 16.07 $Gy_{[10]}$.

If uncorrected the tumour BED (over 31 + 7 days) would be

$$= 23 \times 1.82\ (1.15 + 1.05^2 \times 1.82/10) - 0.3 \times 38 = 45.14\ Gy_{[10]}.$$

By treating through one weekend, two days can be saved, so the tumour BED is then

$$= 23 \times 1.82(1.15 + 1.05^2 \times 1.82/10) - 0.3 \times 36 = 45.74\ Gy_{[10]}.$$

By treating twice a day for all remaining weekdays treatments (i.e. eight fractions in four days) the overall time is 32 days and the BED is then

$$= 23 \times 1.82(1.15 + 1.05^2 \times 1.82/10) - 0.3 \times 32 = 46.94\ Gy_{[10]}.$$

This seems a good alternative but may not be feasible if twice-daily sedation is unacceptable.

Another possibility would be to give additional daily fractions, extending overall time further. If, for example, we try to give the remaining compensatory dose in one fraction in one extra day after completing the prescribed course (and the delay), we obtain an overall treatment time of around 39 days, then

$$\text{BED} = 24 \times 1.82(1.15 + 1.05^2 \times 1.82/10) - 0.3 \times 39 = 47.30\ \text{Gy}_{[10]}.$$

which is almost the same tumour BED as the original prescription, but the NT BEDs will then be tissue A BED = 81.63 $\text{Gy}_{[3]}$ and tissue B BED = 100.61 $\text{Gy}_{[2]}$. These have increased. This additional NT BED can be reduced by fractionating the final days treatment into say two treatments of 1 Gy (physical dose), so that:

the tumour BED is then 47.04 $\text{Gy}_{[10]}$; tissue A BED = 80.89 $\text{Gy}_{[3]}$ and tissue B BED = 99.42 $\text{Gy}_{[2]}$.
This appears to be a reasonable compromise.

Consider what might happen if the remaining treatment is hypofractionated and given in only three fractions in order to complete in the same overall treatment time but matched to give the same CNS effect in tissue B (a deficit BED of 33.53 $\text{Gy}_{[2]}$).

First, solve $33.53 = 3 \times d(1.3 + 1.05^2\ d/2)$; $d = 3.47$ Gy.
Then, the tumour BED would be low at:

$$\text{BED} = 15 \times 1.82(1.15 + 1.05^2 \times 1.82/10) + 3 \times 3.47(1 + 3.47/10) - 31 \times 0.3$$
$$= 43.58\ \text{Gy}_{[10]}.$$

This is probably not acceptable since the BED is less than the originally intended 47.24 $\text{Gy}_{[10]}$. So, the alternative methods given above would be preferred to maintain reasonable tumour control.

Another possibility would be to use a dose such as 0.98 Gy (physical) twice daily for eight days, completing treatment in 36 days.

The tumour BED would be:

$$15 \times 1.82(1.15 + 1.05^2 \times 1.82/10) + 0.98 \times 16(1.15 + 1.05^2 \times 0.98/10) - 0.3$$
$$\times\ 36 = 45.80\ \text{Gy}_{[10]}.$$

The tissue A BED is then:

$$15 \times 1.82(1.2 + 1.05^2 \times 1.82/3) + 0.98 \times 16(1.2 + 1.05^2 \times 0.98/3) = 75.48\ \text{Gy}_{[3]}$$

and the tissue B BED is then:

$$15 \times 1.82(1.3 + 1.05^2 \times 1.82/2) + 0.98 \times 16(1.3 + 1.05^2 \times 0.98/2) = 91.73\ \text{Gy}_{[2]}.$$

Again, incomplete repair between fractions has not been used in the example, but could be used to further refine the BED.

10.2.6 Comparison of the two methods

The fixed and flexible RBE calculations do give different answers when dose per fraction is changed. A comparison for one of the above alternatives is given in table 10.1. Also, in the case of giving three hypo-fractionated doses as compensation, we have BED values of 42.06 $\text{Gy}_{[10]}$ and 43.58 $\text{Gy}_{[10]}$ for the fixed and flexible RBE methods, respectively. These differences—due to the assumed changes in RBE with dose per fraction—provide a good reason why it might be best to deliver the

Table 10.1. BED calculations using two different modelling methods for twice daily fractions of 1.15 Gy-equivalent and 0.98 Gy (physical) on eight treatment days, with treatment completion in 36 days.

	Fixed RBE	Flexible RBE
BED tumour ($Gy_{[10]}$)	45.72	45.80
BED tissue A ($Gy_{[3]}$)	75.45	75.48
BED tissue B ($Gy_{[2]}$)	88.98	91.73

remaining dose mostly using the same fraction size where possible and using altered fractionation on only a few days, especially those days preceding weekends if there is to be a further break in treatment, since over weekends accumulated incomplete repair will not occur. There is a difficulty in using 2 Gy twice daily for many treatments in acute reacting tissues, since reactions can be made worse. General guidelines for limits of acute reacting normal tissues have been given by Fowler *et al* (2003).

Worked example 2

An ion beam schedule, of 64 Gy-eq in 20 fractions over 26 days, is interrupted after 15 fractions, for the entire fourth week and for a further three days into the following week (i.e. for ten days). The tumour being treated is a squamous cell carcinoma of the peripheral bronchial tree. Here we assume a significant change in RBE with dose per fraction and only use the RBE_{max} and RBE_{min} concepts. There is no spinal cord dose. To simplify the calculations we assume that the dose to lung tissue within the PTV will obliterate all lung function within the PTV, also that a surrounding volume (VLNT) for lung function receives approximately 50% of the prescribed dose and is relevant to the patient's future wellbeing.

In this example the following assumptions are made.

RBE for dose conversion = 2.6; tumour RBE_{max} = 3.2; NT RBE_{max} = 4; tumour and NT RBE_{min} = 1.2; tumour α/β = 10 Gy; NT late reacting NT α/β = 3 Gy. Tumour repopulation is modelled by no repopulation for 21 days followed by a repopulation rate equivalent to 0.9 Gy day^{-1}.

The 64 Gy-eq dose is a physical dose of 64/2.6 = 24.62 Gy and the physical dose per fraction consequently is 24.62/20 = 1.23 Gy.

Using the *flexible RBE* model we obtain the following.

The intended tumour BED is given by

$20 \times 1.23(3.2 + \frac{1.2^2 \times 1.23}{10}) - 0.9(T - T_K)$; the final repopulation term is zero since the intended T is less than T_K, so the BED is 83.08 $Gy_{[10]}$.

The intended NT BED is $20 \times 1.23 \times 0.5\left(4 + \frac{1.2^2 \times 1.23 \times 0.5}{3}\right) = 52.83\ Gy_{[3]}$.

The delivered tumour BED is $15 \times 1.23\left(3.2 + \frac{1.2^2 \times 1.23}{10}\right) = 62.31\ Gy_{[10]}$.

The delivered NT BED is $15 \times 1.23 \times 0.5\left(4 + \frac{1.2^2 \times 1.23 \times 0.5}{3}\right) = 39.62\ \mathrm{Gy}_{[3]}$.

The deficit BEDs are then:
tumour deficit: $83.08 - 62.31 = 20.77\ \mathrm{Gy}_{[10]}$,
NT deficit: $52.83 - 39.62 = 13.21\ \mathrm{Gy}_{[3]}$.

If it is intended to proceed by either:

(1) giving the same fraction number (five) over seven days (that is ten days longer than intended), but increasing the total dose,
(2) giving six fractions of a lower dose in three days (that is six days longer) by treating twice daily including on a Saturday.

Then for (1):

(a) if tumour control is to be maintained, we solve:

$$20.77 = 5\,d(3.2 + 1.2^2 d/10) - (36 - 28)0.9$$

$$d = 1.63\ \mathrm{Gy}.$$

The normal tissue BED will then have increased to be:

$39.62 + 0.5 \times 5 \times 1.63\,(4 + 1.2^2 \times 1.63 \times 0.5/3) = 57.52\ \mathrm{Gy}_{[3]}$, which has increased from $52.83\ \mathrm{Gy}_{[3]}$. This could amount to 5%–10% additional complications in the volume of concern if complications increase at 1%–2% per Gy. If such a policy were to be followed, it might be appropriate to reduce the dose per fraction to, say, 1.5 Gy. This would then give a compromised tumour BED of:

$62.31 + 5 \times 1.5\,(4 + 1.2^2 \times 1.5/3) - (36 - 28)\,0.9 = 80.73\ \mathrm{Gy}_{[10]}$ instead of the intended $83.1\ \mathrm{Gy}_{[10]}$.

Ultimately, it is the oncologist who has to decide which policy to pursue and alter the radiation prescription but since the risks and benefits may have changed, the patient should be informed as to the decision made and, ideally, they should be consulted when it comes to the choice of risks.

For method (2), where six fractions are given in three consecutive days (a total of 32 days), we can maintain the same tumour BED by solving for d in:

$$20.77 = 5d(3.2 + 1.2^2 d/10) - (32 - 28)0.9$$

$$d = 1.2\ \mathrm{Gy}.$$

Then the NT volume BED increases from the intended $52.83\ \mathrm{Gy}_{[3]}$ to be

$$39.62 + 0.5 \times 5 \times 1.2(4 + 1.2^2 \times 1.2 \times 0.5/3) = 55.06\ \mathrm{Gy}_{[3]}.$$

The latter option appears more acceptable than the former option (1), although we have not included a small increment for incomplete repair between fractions, which is considered to be small for inter-fraction intervals of 12 hours for only six fractions.

This exercise has assumed that the RBE operative in the normal tissue is a 'worst-case' scenario; at an increasing distance from the target the LET will fall as will the RBE. However, with spread-out Bragg peaks the LET remains quite high. Such details are more readily available using specially adapted treatment planning systems, but which are beyond the scope of the present article. Again, as in the proton case discussed above, a lower dose per fraction could be chosen, e.g. 1.15 or 1.18 Gy. In all cases of a lower dose the tumour BED would fall below that intended originally.

A further example of an ion beam boost following a first phase of x-ray based treatment can be found in Jones *et al* (2006).

It must be remembered that there are three overall approaches to obtain a practical solution in these treatment interruption situations. These are the following.

1. Treat with same dose per fraction, but extend the overall treatment time, possibly increasing the number of fractions. This at least has the potential advantage that the RBE will not change.
2. Treat with larger doses per fraction and complete in the same overall treatment time. In this case the RBE will decrease since fraction size has increased.
3. Treat with a larger number of fractions, but with smaller dose per fraction over a slightly extended treatment time or the same treatment time. In this case the RBE will increase since the fraction size is smaller. This may have a larger effect in the normal tissues since the maximum RBE is greater in late reacting tissues.

The above examples are meant to serve as a demonstration of how to approach treatment interruption corrections in charged particle therapy for treatments of curative (radical) intent. Similar approaches, with some modifications, can be applied to palliative treatments where the endpoint is the relief of symptoms for a reasonable duration.

10.3 Summary for unintended treatment gap corrections

- It can be appreciated that an individual approach should be used, although formulaic methods can also be devised for precise compensation for either the tumour or the normal tissue iso-effect. However, compromise solutions are inevitably required and individualisation is consequently essential.
- Ideally, such corrections should be done by persons who will practice and become used to these approaches, as there are many potential pitfalls.
- It has even been suggested that nation states should organise their own referral system for doing such calculations, although worldwide solutions and more global governance are feasible alternatives in an electronic data age.
- Data collection and sharing should be encouraged in order to study and gain further useful information.

10.4 Re-treatments

10.4.1 Background

Radiotherapy retreatment can be of benefit in carefully selected patients with recurrent local disease and where other salvage treatment options are inappropriate (Stewart 1999). Methods for estimating the change in spinal cord radiation tolerance with increasing time between the first and second radiotherapy courses have been published by Jones and Grant (2014), and Jones and Hopewell (2014). The mechanism of effect is late vascular damage and although it is possible that similar kinetics apply to late vascular damage in other organ systems, this is not necessarily the case; there may be anomalous tissue such as the kidney where late damage is amplified by hypertension caused by the initial renal damage itself, causing a vicious cycle of damage and so the same relationships should not be applied. However, most late-reacting tissues, including the CNS, have similar time courses for late damage expression, although the α/β ratio used may differ. In this sense we apply more conservatism on CNS sub-organs at risk than in many other anatomical sites by using $\alpha/\beta = 2$ Gy, rather than 3 Gy elsewhere. There are also some unsatisfactory aspects to many of the classical small animal re-treatment experiments where beams also irradiated adjacent tissues or even the hemi-body. Also, for example, bladder irradiations included the whole bladder rather than a part of it so the late effect studied was centripetal constriction resulting in urinary frequency. Such a late effect is as severe after irradiating only a part of the organ.

These models rely on the basic linear-quadratic (LQ) concept of BED and some generic conclusions that after moderate to high doses of radiation there is around 50% recovery of tolerance in the spinal cord at around six months following radiotherapy. The study by Jones and Hopewell (2014) was based on the primate spinal cord study of Ang *et al* (2001), where an initial fixed dose was followed by retreatment at time intervals of one, two and three years. Essentially, if the first and second treatment courses are expressed as their BEDs as a percentage of their tolerance BED, i.e. (given BED/tolerance BED)%, then it is possible to apply a time-dependent function, $r(t)$, to estimate the final percentage tolerance BED from the first. It is important to carefully follow these definitions:

BED_{init} is the initial course BED, BED_{final} is the final course BED;
BED_1 is the % tolerance BED for the initial course, (BED_{init}/tolerance BED)%;
BED_2 is the % tolerance BED for the final course, (BED_{final}/tolerance BED)%.

From the shape of the experimental BED_1 versus BED_2 plots, a reasonable model, which can be fitted to the animal data, is

$$BED_2 = 100\left(1 - \frac{BED_1}{100}\right)^{\frac{1}{r(t)+1}} \tag{10.6}$$

where $r(t)$ is expressed as $r(t) = 1.5 + \exp[1.2\,(t - 1)]$ in a conservative fit of the data, with t in units of years.

Thus if BED_{init} and the tolerance BED are known, then for r = 1–3 years BED_2 can be estimated. From BED_2, BED_{final} can be found. For retreatments six months after previous exposures for BED_1 values of over 30%, it can be more simply assumed that 50% recovery of BED occurs since many rat studies have shown such a figure. This approach would require a calculation based on:

$$\text{available re-treatment BED} = \text{tolerance BED} - 0.5\ BED_{init}.$$

Caution is required between six months and one year when an intermediate value between the simple method (at six months) and the equation based method described above (at one year), or if in doubt keep to the six months recovery level. In fact if an interval of 0.5 years is assumed within equation (10.1) and the overall recovery is reasonably close to the animal data, equation (10.1) could be used with caution for $r < 1$ year, but the user should be aware that the approach falls outside the rhesus monkey data range. Extreme caution is required if full tolerance doses have already been delivered; the methods described are valid only for sub-tolerance doses.

For charged particles, it is necessary to use the same BED approach, where the particle BEDs are expressed as a function of α/β, dose, number of fractions, RBE_{max} and RBE_{min}. The latter two parameters can be estimated from the LET.

It could be the case that the initial treatment has used megavoltage photons and the second treatment charged particles. The appropriate form of the BEDs must be used in each case, for the prescription and the chosen tolerance level. Note that the chosen tolerance level should be the same in the first and second treatment courses, as this was the case in the experimental data analysis.

Care is required after BED_2 has been estimated, not only to convert BED_2 to BED_{final}, but also to determine the charged particle physical dose.

A worked example is given, where for educational purposes it is assumed that both initial and final treatments were given by charged particles.

Worked example 3

A low-grade tumour has been treated with protons to 50 Gy-RBE in 25 fractions (using an overall RBE of 1.1), the surrounding (brain) tolerance being 60 Gy in 30 fractions (megavoltage photons). Three years later, the tumour has recurred outside one tumour edge and it is decided to re-treat the tumour and some areas of the brain that overlap into the previous volume. Determine a safe re-treatment dose using protons. (Here it is assumed that the LET has changed for each treatment, such that the $RBE_{max} = 1.4$ and RBE_{min} is 1.05 in both courses; $(\alpha/\beta)_L$ = 2 Gy.)

Calculations

It is important to remember that the 50 Gy-RBE in 25 fractions is a physical dose of 50/1.1= 45.45 Gy and a dose per fraction of 2/1.1 = 1.82 Gy.

The high LET BED (equation (2.16) in chapter 2) is

$$\mathrm{BED} = \boldsymbol{D}_{\mathrm{H}}\left(\mathrm{RBE}_{\max} + \frac{\mathrm{RBE}_{\min}^{2}\, d_{\mathrm{H}}}{\left(\frac{\alpha}{\beta}\right)}\right) \tag{10.7}$$

and so,

$$\mathrm{BED}_{\mathrm{init}} = 45.45(1.4 + 1.05^{2\prime} \times 1.82/2) = 109.23\ \mathrm{Gy}_{[2]}.$$

The working tolerance BED is 60(1 + 2/2) = 120 $\mathrm{Gy}_{[2]}$,
so, BED_1 = (109.23/120) × 100 = 91.03%.

Next $r(t)$ is obtained from $r = 1.5 + \exp[1.2\,(3 - 1)]$, where the elapsed time is three years.

From which $r(t) = 12.52$.

Then using equation (10.1) given above,

$$\mathrm{BED}_2 = 100\left(1 - \frac{91.03}{100}\right)^{\frac{1}{12.52+1}} = 83.67\%.$$

Then find the safe $\mathrm{BED}_{\mathrm{final}}$, which is contained in the definition of BED_2 as

$\mathrm{BED}_2 = (\mathrm{BED}_{\mathrm{final}}/120) \times 100$, and so $\mathrm{BED}_{\mathrm{final}} = \mathrm{BED}_2 \times 120/100$. Then for a BED_2 of 83.67% (rounded up), the $\mathrm{BED}_{\mathrm{final}}$ is 100.4 $\mathrm{Gy}_{[2]}$.

To obtain this BED, the high LET BED equation is used again, but this time with different RBE parameters owing to the change in LET, and if the retreatment schedule is given in 20 fractions then

$$100.4 = 20d(1.4 + 1.05^2 \times d/2)$$

and the solution for d is 2.0 Gy, so that 20 × 2 = 40 Gy of protons, a physical dose inclusive of the RBE (so this would be 44 Gy-RBE if using the constant RBE of 1.1). Alternatively for 25 fractions the solution of $100.4 = 25\,d(1.4 + 1.05^2 \times d/2)$ provides $d = 1.71$ Gy.

It is interesting to then find the operative RBE for the estimated re-treatment dose of 2 Gy. This can be found by solving for d_{L} in the iso-effect equation for the 20 fraction schedule where $d_{\mathrm{H}} = 2$ Gy:

$$d_{\mathrm{L}}\left(1 + \frac{d_{\mathrm{L}}}{(\alpha/\beta)_{\mathrm{L}}}\right) = 2\left(1.4 + \frac{1.05^2 \times 2}{(\alpha/\beta)_{\mathrm{L}}}\right)$$

where the positive value for d_{L} is rounded to 2.32 Gy and so the RBE will be, by definition, $d_{\mathrm{L}}/d_{\mathrm{H}}$ or 2.32/2 = 1.16, approximately. For the alternative 25 fraction schedule where $d_{\mathrm{H}} = 1.71$ Gy, the solution for d_{L} is 2 Gy and the operative RBE is then 2/1.71 = 1.17, a slight increase.

These are conservative estimates and higher doses may be selected, especially for small volumes but if even lower doses should be considered if there are factors that indicate a lower tolerance, and higher risk, because of multiple surgical approaches, the magnitude of cytotoxic chemotherapy exposure and other concurrent medical conditions including age, etc. It is important to establish if the re-treatment is of radical or palliative intent.

For this re-treatment example, it is of further interest to find what the 'safe dose' would be if a 'constant RBE' of 1.1 was used. In this case $RBE_{max} = RBE_{min} = 1$, and further calculations lead to re-treatment physical doses of 43.5 Gy in 20 fractions or 47 Gy in 25 fractions (the Gy-RBE doses using the 1.1 RBE would be 47.85 and 51.7 Gy-RBE, respectively. In each option, higher doses are estimated in comparison to the flexible RBE approach, and so could result in a greater risk of toxicity.

Alternative clinical situations where the first treatment was megavoltage photons and the second by charged particles such as protons, or vice versa, can be handled by the same processes as described above.

Further work is being done on refining the above model for times between 0 and 1 year, and further risk estimation inclusion, and will soon be available in a GUI under the auspices of The Department of Applied Mathematics, Oxford.

Just as with treatment interruptions, with re-treatments there are also important extra steps within the calculations when high LET radiations are used. Considerable care is required.

References

Ang K K, Jiang G L, Feng Y, Stephens L C, Tucker S L and Price R E 2001 Extent and kinetics of recovery of occult spinal cord injury *Int. J. Radiat. Oncol. Biol. Phys.* **50** 1013–20

Bese N S, Hendry J and Jeremic B 2007 Effects of prolongation of overall treatment time due to unplanned interruptions during radiotherapy of different tumor sites and practical methods for compensation *Int. J. Radiat. Oncol. Biol. Phys.* **68** 654–61

Dale R G *et al* 2002 Practical methods for compensating for missed treatment days in radiotherapy, with particular reference to head & neck schedules *Clin. Oncol.* **14** 382–93

Fowler J F, Denekamp J, Sheldon P W, Smith A M, Begg A C, Harris S R and Page A L 1974 Optimum fractionation in x-ray treatment of C3H mouse mammary tumours *Brit. J. Radiol.* **47** 781–9

Fowler J F, Harrari P M, Leborgne F and Leborgne J H 2003 Acute radiation reactions in oral and pharyngeal mucosa: tolerable levels in altered fractionation schedules *Radiother. Oncol.* **69** 161–8

Jones B, Carabe-Fernandez A and Dale R G 2006 Calculation of high-LET radiotherapy dose required for compensation of overall treatment time extensions *Brit. J. Radiol.* **79** 254–7

Jones B and Grant W 2014 Retreatment of central nervous system tumours *Clin. Oncol.* **26** 407–18

Jones B and Hopewell J H 2014 Alternative models for estimating the radiotherapy retreatment dose for the spinal cord *Int. J. Radiat. Biol.* **90** 731–41

Kim J J and Tannock I F 2005 Repopulation of cancer cells during therapy: an important cause of treatment failure *Nature Rev. Cancer* **5** 516–25

Royal College of Radiologists 2008 *The Timely Delivery of Radical Radiotherapy: Standards and Guidelines for the Management of Unscheduled Treatment Interruptions* 3rd edn 2008 (London: Royal College of Radiologists)

Stewart F A 1999 Re-treatment after full-course radiotherapy: is it a viable option? *Acta Oncol.* **38** 855–62

Thames H D *et al* 2010 The role of overall treatment time in the outcome of radiotherapy of prostate cancer: an analysis of biochemical failure in 4839 men treated between 1987 and 1995 *Radiother. Oncol.* **96** 6–12

Withers H R, Taylor J M and Maciejewski B 1988 The hazard of accelerated tumor clonogen repopulation during radiotherapy *Acta Oncol.* **27** 131–46

IOP Publishing

Practical Radiobiology for Proton Therapy Planning

Bleddyn Jones

Chapter 11

Errors of Bragg peak positioning and their radiobiological correction

11.1 Introduction

In charged particle therapy, using protons and heavier ions, energy deposition occurs in an entirely different way from the x-rays/photons used in conventional radiotherapy. Particle Bragg peaks provide selective dose distributions, reducing collateral radiation to many important tissues and so improve the results of radiotherapy. For this to be fully realised, it is essential to overcome two inherent difficulties associated with particle therapy.

1. Optimal prediction and confirmation of Bragg peak position within the body.
2. Improved knowledge of the relationship between the increased ionisation density within Bragg peaks (quantified as linear energy transfer (LET)) and biological effects. This is necessary in order to prescribe the best possible tumour dose and restrict critical normal tissue doses to the safest possible extent.

It is important not to consider these factors as separate, since errors of Bragg peak positioning will lead to altered dose and LET, with consequent nonlinear effects on biological or clinical outcomes.

This chapter uses the new simpler model for predicting the relative biological effect (RBE) from LET within biological effective dose (BED) equations, as given in chapters 8 and 9. The latter are useful in clinical practice and have been adapted for particle therapy. The new model requires only input of the low LET α and β parameters and the Z number (nuclear charge) of any ion and is independent of further micro-dosimetry considerations, but relies on saturation and efficiency concepts. It can therefore be used as a second check for more complex models used at the present time, and for tentative radiobiological modelling. The model fits

doi:10.1088/978-0-7503-1338-4ch11

data from many classical particle beam experiments using different ions and at different doses or levels of cell survival.

Modelling the effect of erroneous dose delivery shows that particle therapies are more susceptible to a loss of tumour control than are x-ray based treatments. Thus particle therapy demands greater attention to detail with best use of image guidance techniques. The importance of using the most correct RBE is emphasised, since if the RBE is incorrect, there is increased susceptibility to treatment delivery errors.

A method for compensating unintended dose delivery errors, with worked examples, is presented. Changes in dose and LET (with resultant changes in radio-sensitivity α and β values) are fed into BED equations that are commonly used to correct errors in photon treatments. To implement these would be time consuming since 3D LET mapping of patient imaging data is required; this is not routinely available at the present time. In principle, the method provides a rational system for achieving outcomes close to what was originally intended.

Some further definitions for the terminology of the suggested corrective measures are given here.

11.1.1 Further abbreviations and definitions

ForLET (assuming LET is an appropriate average)

$_{\mathrm{T}}L_{\mathrm{I}}$ = intended tumour LET
$_{\mathrm{T}}L_{\mathrm{E}}$ = erroneous tumour LET
$_{\mathrm{N}}L_{\mathrm{I}}$ = intended normal tissue LET
$_{\mathrm{N}}L_{\mathrm{E}}$ = erroneous normal tissue LET
$_{\mathrm{T}}L_{\mathrm{C}}$ = tumour LET during corrective treatments
$_{\mathrm{N}}L_{\mathrm{C}}$ = normal tissue LET during corrective treatments.

Fordose, which is assumed to be the physical dose and not the RBE-corrected dose at this stage:

$_{\mathrm{T}}d_{\mathrm{I}}$ = intended tumour dose
$_{\mathrm{T}}d_{\mathrm{E}}$ = erroneous tumour dose
$_{\mathrm{N}}d_{\mathrm{I}}$ = intended normal tissue
$_{\mathrm{N}}d_{\mathrm{E}}$ = erroneous normal tissue dose.

11.1.2 Background considerations

Delivery of radiotherapy treatment may be inaccurate and unreliable even in the case of conventional (x-ray or photon)-based radiotherapy, where systematic and random errors can occur for a wide variety of reasons. Incorrect dose may be due to accelerator output errors, dose calculation errors for complex treatments where dose is estimated using computer algorithms that may not necessarily take into account the true complexity of intervening tissue in-homogeneities, the effects of patient/tumour movement with time during and between treatment fractions, and other temporal changes in tumour or patient anatomy during a fractionated treatment course. This error proneness is potentially greater for particle therapy—the dose calculation may not be as reliable due to other factors inherent in the total system of treatment dose planning, including variations in beam delivery and particle range differences superimposed on the photon situation, which can result in marked changes in the intended highly selective energy deposition. This energy, when locally absorbed, is quantified in two ways: as LET on a microscopic scale, and as dose in the conventional way. The added difficulty is that LET and dose both influence the relative biological effect (or efficiency), often referred to as the RBE. The RBE is not required in the calculation of dose correction for errors made in conventional photon therapy, but changes in dose, and also in some instances LET, will each influence the 'correct dose' for a patient, not only in terms of tumour dose but also dose to the critical normal tissues.

The high LET radiation BED equations, which contain the RBE limits termed RBE_{max} and RBE_{min}, can be used to estimate iso-effective fractionation schedules as well as to correct unintended treatment interruptions (Jones *et al* 2006). BED equations have been recommended as the method of choice to calculate dose corrections after errors of photon based radiotherapy delivery (Jones and Dale 2008), but no general system has been described for calculation of treatment delivery errors when using positively charged particle beams. The purpose of this report is to assess the potential impact of such errors and also describe how to compensate for them.

11.2 Brief description of methods

The radiobiological linkage between LET, Dose and RBE is based on the efficiency model already given in chapters 8 and 9. Any error in LET and dose will produce a deviancy of ΔL (for LET) and Δd for dose. For simplification here we assume a precise value of LET rather than in more practical situations where a spread of LET values over a specified range of energies is inevitably used in the treatment delivery process, and which would complicate matters further.

The following approach briefly describes how to determine the change in BED due to any error. It is necessary to first assume reasonable values of the low LET α and β parameters.

1. Estimate the effect of any change in LET initially and its effect on α and β. This can be achieved quite simply from the particle Z number, which determines LET_U.
2. Determine, from the changes in α and β, the changes in RBE_{max} and RBE_{min} and use these to estimate the change in BED due to the error.
3. From a new treatment plan, determine if LET is changed as well as dose. Use the same equations to determine any changes in α, β and so in RBE_{max} and RBE_{min}.
4. Use these values of RBE_{max} and RBE_{min} within the BED equations to determine the dose required for maintaining the required tumour BED.

Re-iteration of these final steps may be required to produce a final satisfactory or adequate compromise solution.

This overall procedure will be described in greater detail below. It will be important to compare such a novel and relatively simple approach with those obtained by assuming the more complex micro-dosimetry systems quoted in chapter 8.

11.3 Description of the model

11.3.1 BED equations

The BED (sometimes referred to as EQD_0) is normally, for low LET (1–1.5 keV μm^{-1}) radiations:

$$\mathrm{BED} = D_\mathrm{L}\left(1+\frac{d_\mathrm{L}}{\left(\frac{\alpha}{\beta}\right)}\right) \tag{11.1}$$

where D_L is the total dose and d_L the dose per fraction, the two being linked by the number of fractions N, where $Nd_\mathrm{L} = D_\mathrm{L}$, and α/β is a biological parameter which controls fractionation sensitivity for the control low LET radiation, and has units of dose. For higher values of LET the equation changes to be:

$$\mathrm{BED} = D_\mathrm{H}\left(\mathrm{RBE}_{\max} + \frac{\mathrm{RBE}_{\min}^2 d_\mathrm{H}}{\left(\frac{\alpha}{\beta}\right)}\right). \tag{11.2}$$

In this last equation, it is important to note that the α/β ratio here remains unchanged from that of the control low LET radiation. The low LET α/β ratio is effectively acting as a constant, but is modified by the presence of the two limiting RBE parameters.

Next, it is essential to link the RBE values to the prevailing LET value. This is necessary because the relationship between LET and RBE, although linear initially (when plotted on linear scales) becomes nonlinear, with a turnover point at a critical value of LET (termed $\mathrm{LET_U}$), after which the RBE decreases with LET down to a value of just above 1 in the case of beams used in medical applications. However, for these applications it is rare to experience LET values greater than the turnover point since treatment volumes are covered by a mixture of LET values from parts of individual particle beams, which have lower LET values proximal to the Bragg peak and higher values within, and especially towards the end of, the peaks.

The method follows some simple steps, each with numerical parameters given in chapters 8 and 9. In brief:

1. $\mathrm{LET_U}$ is obtained from $\mathrm{LET_U} = 30.5 + S/k\,(1 - \exp[-k\,(z - 1))$ ($S = 114.4$, $k = 0.51$).
2. α_U is found from $\alpha_\mathrm{U} = A/j\,(1 - \exp[-j\,\alpha_\mathrm{L})$ ($A = 10.6$, $j = 3.9$).
3. β_U is found from $\beta_\mathrm{U} = B/i\,(1 - \exp[-i\,\beta_\mathrm{L})$ ($B = 3.2$, $i = 40$).
4. Values of α_H and β_H at any particular value of LET (LET_x) are found from the following.

(a) A relatively simple equation representing efficiency changes between the control (megavoltage photon) radiation LET value and the LET_U value, as a change in radiosensitivity from α_L to be a larger value of α (α_H) at any higher value of LET (LET_x) up to a maximum possible value of α given by α_U. The same LET_U is also assumed for β (in order to preserve symmetry with dose). So, we obtain:

$$\alpha_H = \alpha_C + \frac{LET_x - LET_C}{LET_U - LETc}(\alpha_U - \alpha_C)$$

$$\beta_H = \beta_C + \frac{LET_x - LET_C}{LET_U - LETc}(\beta_U - \beta_C)$$

(b) A simple equation representing inefficiency for all LET values above LET_U, but converted into an efficiency equation, as:

$$\alpha_H = \alpha_C + \left(1 - \frac{LET_x - LET_C}{LET_U}\right)(\alpha_U - \alpha_C).$$

$$\beta_H = \beta_C + \left(1 - \frac{LET_x - LET_C}{LET_U}\right)(\beta_U - \beta_C).$$

5. The RBE_{max} and RBE_{min} values can then be obtained.
6. For the RBE, corrected for dose per fraction, it is necessary to solve an iso-effect equation where $\alpha_L d_L + \beta_L d_L^2 = \alpha_H d_H + \beta_H d_H^2$ is solved for d_L, which is then divided by the d_H to give the RBE at any value of d_H. In terms of BED, this means solving the iso-effect equation BED of control radiation = BED of particle radiation, i.e.

$$nd_L\left(1 + \frac{d_L}{k}\right) = nd_H\left(RBE_{max} + \frac{RBE_{min}^2 d_H}{k}\right)$$

the RBE is then the solution of this equation for d_L, then divided by d_H.

Examples of this method to obtain RBE, fitted to various data sets, which include changes in dose (or surviving fraction), have been given in chapters 8 and 9.

11.3.2 Assessment of BED changes after an error

From estimates of the changes in α and β described above, the operative RBE_{max} and RBE_{min} in equation (11.2) above can be determined. The intended, delivered and corrected BEDs can then be calculated providing the LET values and dose for these states are known. Any excess or deficit in BED is then obtained as the difference between the intended and delivered BEDs in both tumour and normal tissue. Any unintended over-dosage or under-dosage can then be compensated for by a new, modified dose to provide the originally intended tumour BED.

The above system applies to mono-energetic particles, or a small range of particle energies, which is reasonable for modelling purposes at this stage and could be extended to ranges if LET in accordance with energy ranges, by an averaging process.

With the above relatively simple-to-operate method for predicting changes in RBE due to changes in both LET and dose, it becomes possible to tentatively assess the clinical impact of changes in Bragg peak positioning during particle therapy, and for any ion species. For example, the intended BED at a LET of ${}_{\mathrm{T}}L_{\mathrm{I}}$ will be given by:

$$n\ {}_{\mathrm{T}}d_{\mathrm{I}}\ (\mathrm{RBE}_{\mathrm{max}}+\ \mathrm{RBE}^{2}_{\mathrm{min}}\ {}_{\mathrm{T}}d_{\mathrm{I}}/(\alpha/\beta)_{\mathrm{L}} = n\ {}_{\mathrm{T}}d_{\mathrm{I}}\ (\alpha_{\mathrm{H}}/\alpha_{\mathrm{L}}+\ (\alpha_{\mathrm{H}}/\beta_{\mathrm{L}})\ {}_{\mathrm{T}}d_{\mathrm{I}}/(\alpha/\beta)_{\mathrm{L}})$$

and the values of α_{H} and β_{H} for the operative LET can then be obtained from the steps 1–5 in the list given in section 11.3.1.

For a different dose ${}_{\mathrm{T}}d_{\mathrm{E}}$ at a different LET value ${}_{\mathrm{T}}L_{\mathrm{E}}$, the BED for n fractions is then:

$$n\ {}_{\mathrm{T}}d_{\mathrm{E}}(\alpha_{\mathrm{H}}/\alpha_{\mathrm{L}}+\ (\alpha\mathrm{H}/\beta\mathrm{L})\ {}_{\mathrm{T}}d_{\mathrm{E}}/(\alpha/\beta)_{\mathrm{L}}),$$

which is the same as

$$n\ {}_{\mathrm{T}}d_{\mathrm{E}}\ (\mathrm{RBE}_{\mathrm{max}}+\ \mathrm{RBE}^{2}_{\mathrm{min}}\ {}_{\mathrm{T}}d_{\mathrm{E}}/(\alpha/\beta)_{\mathrm{L}}).$$

For example, consider protons, helium ions and carbon ions for an under-dose of 50% of the intended dose. The plots shown in figures 11.1(a)–(f) show estimated changes in BED per fraction for a range of LET values up to the turnover point of the LET–RBE relationship. It should be noted that the lower range of these values pertain to clinical treatments, due to overlapping of Bragg peak and non-Bragg peak regions of the beams. It should also be noted that the highest BEDs are found for the lowest α_{L} radiosensitivity parameter values, because of the greater relative change in α when α_{L} is small, which causes higher $\mathrm{RBE}_{\mathrm{max}}$ values.

An 'ICRU' matrix for particle beams has already been presented in chapter 4. This system should be borne in mind, since shifts in dose (and LET) from one ICRU sub-volume to another will be considered below.

11.3.3 Simulations of treatment delivery errors by modelling

Two simple modelling exercises, each for a population of 100 individuals, are shown below in figures 11.2(a) and (b). These show dose response curves for tumour control probability (TCP) with increasing number of under-dosed fractions ranging from 3 to 15 for intended 2 Gy photon fractions (solid lines) and the same for carbon ions (dashed lines). Figure 11.2(a) shows the situation where the same part of the tumour (10% of its volume) is under-dosed to 0.6 Gy rather than 2 Gy (these doses being the physical doses used for both photons and carbon ions for simplicity). It can be seen that the fall in TCP with increasing number of erroneous fractions is steeper for carbon ions than for photons. In figure 11.2(b), random under-dosage is allowed,

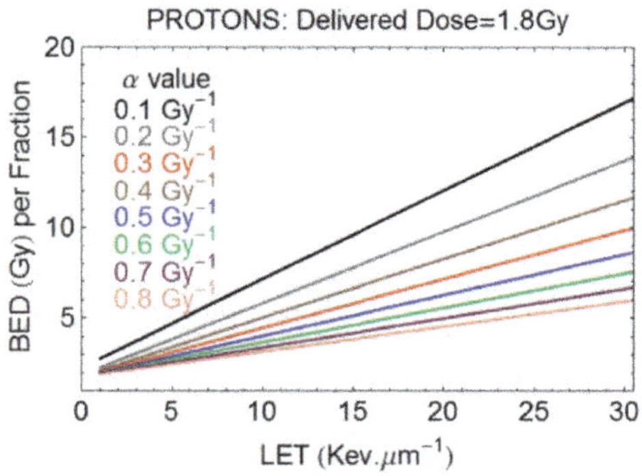

Protons for 1.8 Gy.

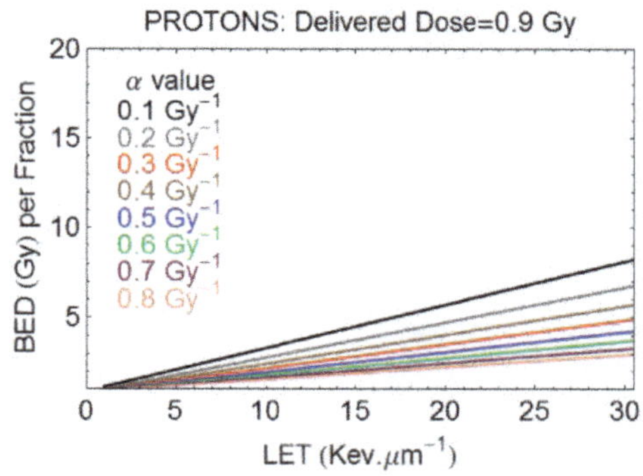

Protons for 0.9 Gy.

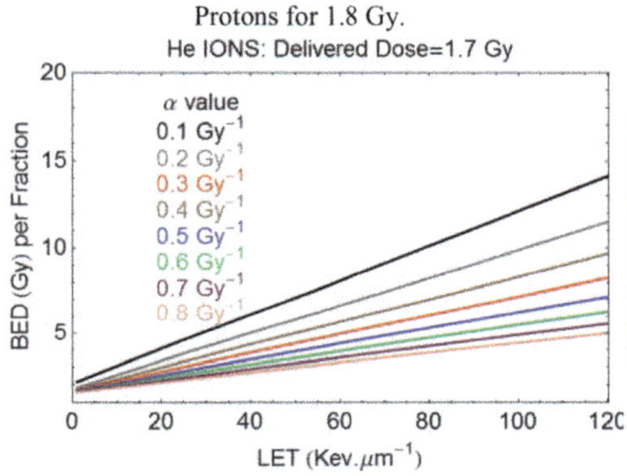

Helium ions at dose of 1.7 Gy

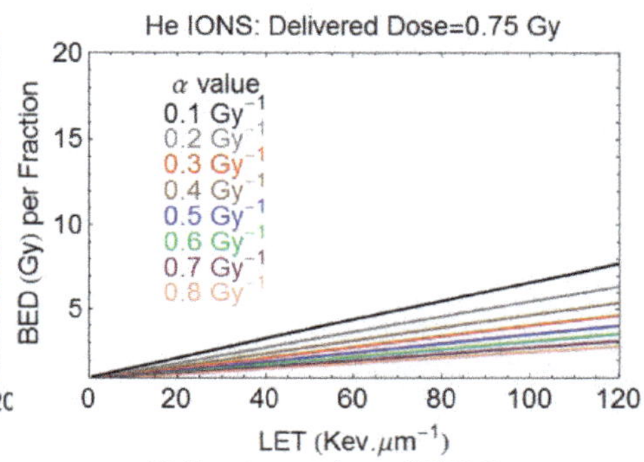

Helium ions at dose of 0.75 Gy

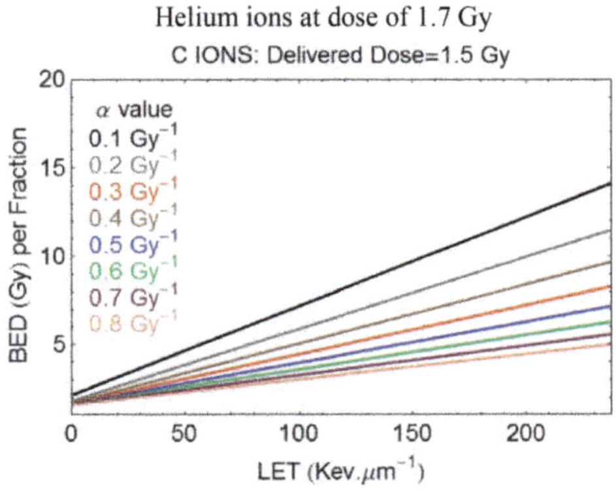

Carbon ions at dose of 1.5 Gy

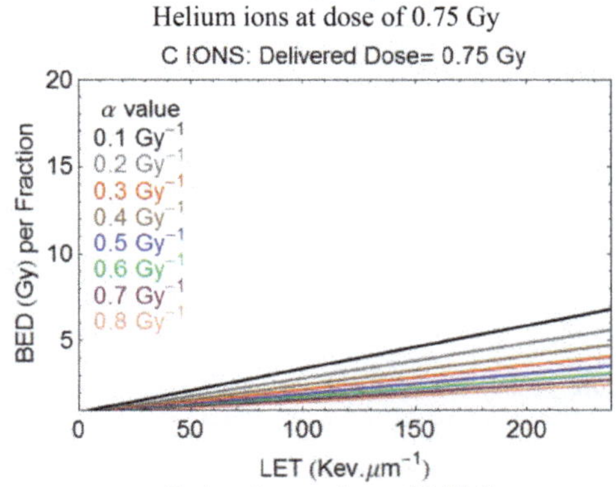

Carbon ions at dose of 0.75 Gy

Figure 11.1. Plots (a)–(f) of BEDs for different values of LET for different ionic species at two values of relatively low doses, with variations in the α_L radiosensitivity parameter, which is labelled as α in the graphs; β is kept constant in this study.

where 40% of the tumour volume is under-dosed to the same amount. The fall-off in TCP is not so significant in this case for both types of radiation, although the slope is again steeper for carbon. The steeper slope is due to a greater proportion of the curative dose being given in fewer fractions with C ions, so in such cases it is more important to avoid under-dosage.

These two examples have not considered changes in LET. Changes in dose, LET and RBE are considered further below.

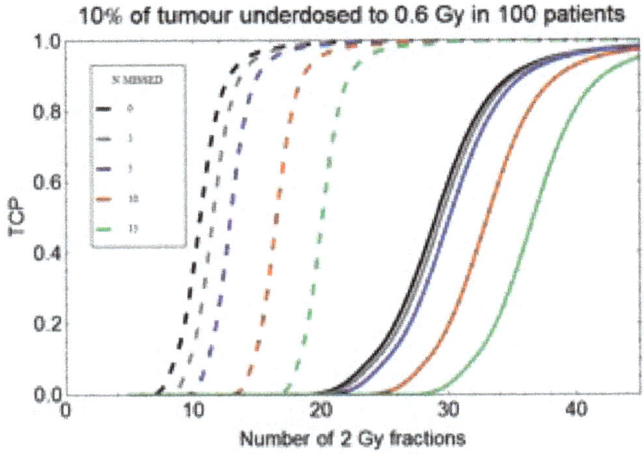

(a)

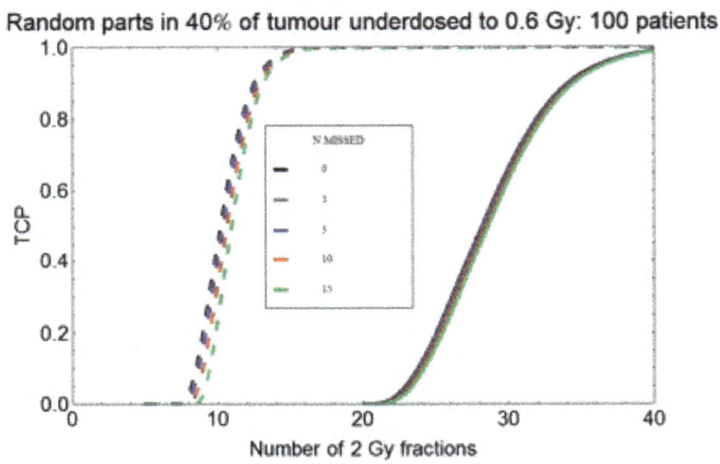

(b)

Figure 11.2. (a) and (b) Plot of TCP against number of 2 Gy fractions for variable numbers of erroneous treatments (solid curves photons, dashed curves carbon ions), where the same 10% of the tumour volume is randomly under-dosed. Assumptions made: RBE_{max} = 3, RBE_{min} = 1.25, using random sampling for the α radiosensitivity parameter (mean 0.3 Gy^{-1}, with standard deviation 0.05 Gy^{-1}) for a tumour with 5 × 108 cells and a constant value of β = 0.03 Gy^{-2} (repopulation not included).

11.3.4 Worked examples of errors and their correction

Next some quantitative, but simplified, worked examples will be considered.

Worked example A

Half of the volume of a tumour is under-dosed by 50% for three fractions, during which the average LET falls from 45 to 15 KeV μM. The prescribed tumour dose was 70 Gy-Eq in 15 fractions using carbon ions.

1. A squamous cell carcinoma where for megavoltage photons $\alpha = 0.35$ Gy^{-1}, $\beta = 0.04$ Gy^{-2}, and the prescription RBE was 2.5.

2. A prostate cancer where for megavoltage photons $\alpha = 0.18\ Gy^{-1}$, $\beta = 0.035\ Gy^{-2}$, and the prescription RBE was 2.

Calculate compensatory doses to the under-dosed half of the tumour for the remainder of treatment in order to maintain iso-effective tumour control.

Worked example B

What is the potential impact on a critical volume of normal tissue planned to receive a dose of 40 Gy in 15 fractions, with the assumption of an operative RBE = 2, which in the past has represented tolerance. Then, for three fractions the dose is erroneously increased by 50% with an increment in average LET from say 15 to 45 KeV μM^{-1}, assuming a 1% increase in serious complications per unit rise in BED above tolerance, and where for megavoltage photons $\alpha = 0.1\ Gy^{-1}$, $\beta = 0.03\ Gy^{-2}$?

Answers

The RBE_{max} and RBE_{min} can be found from the changes in α and β parameters with LET, which can then be used within BED equations together with the low LET α/β value. An ultimate LET_U value of around 230 keV μm^{-1} is used for C ions, compatible with the data of Weyrather *et al* (1999) and the linear relationship between LET, α and β for values less than LET_U.

Results of calculations, using equations given in section 11.3.1, provide the changes in radiosensitivity parameters, RBEmax and RBEmin for the different LET values are shown in table 11.1. These values are placed within the BED equations below where appropriate.

Note that the intended physical dose per fraction was 70/15 = 4.667 Gy-eq which when divided by the assumed RBE provides a physical dose of 4.667/2.5 = 1.866 Gy in example A1, but 4.667/2 = 2.33 Gy for example A2.

Table 11.1. Parameters for worked examples A1 and A2.

	A1: SCC	A2: Prostate
	($\alpha = 0.35$, $\beta = 0.040$)	($\alpha = 0.18$, $\beta = 0.035$)
At LET = 45	$\alpha_U = 2.035$	$\alpha_U = 1.402$
	$\alpha_H = 0.662$	$\alpha_H = 0.406$
	$RBE_{max} = 1.892$	$RBE_{max} = 2.258$
	$\alpha_U = 0.0555$	$\alpha_U = 0.045$
	$\beta_H = 0.0429$	$\beta_H = 0.037$
	$RBE_{min} = 1.035$	$RBE_{min} = 1.027$
At LET = 15	$\alpha_H = 0.454$	$\alpha_H = 0.255$
	$R_{max} = 1.297$	$R_{max} = 1.419$
	$\beta_H = 0.041$	$\beta_H = 0.0359$
	$RBE_{min} = 1.012$	$RBE_{min} = 1.0128$
At LET = 53	$\alpha_H = 0.780$	$\alpha_H = 0.447$
	$RBE_{max} = 2.051$	$RBE_{max} = 2.482$
	$\beta_H = 0.434$	$\beta_H = 0.0372$
	$RBE_{min} = 1.04$	$RBE_{min} = 1.031$

For worked example A1, proceed as follows, with $\alpha/\beta = 8.75$ Gy.
The intended tumour BED, with a dose per fraction of 70/2.5/15 = 1.867 Gy was

$$70/2.5(1.892 + 1.035^2 1.867/8.75) = 59.375\ \text{Gy}_{[8.75]}.$$

The BED after the three fraction under-dosage to the affected part of the tumour was

$$3 \times 0.778\ (1.297 + 1.012^2\ 0.778/8.75) = 3.855\ \text{Gy}_{[8.75]}.$$

This means that 59.375−3.855 = 55.488 Gy is required from the remaining 12 treatments. The dose per fraction is then obtained from the solution of:

$$d\,(2.051 + 1.041^2 d/8.75).$$

This gives $d = 2.01$ Gy.

For worked example A2. The intended tumour BED was $70/2(2.26 + 1.027^2\ 2.33/5.14) = 95.76\ \text{Gy}_{[5.14]}$.

The BED given after the three fraction under-dosage was

$$3 \times 2.33 \times 0.5\ (1.42 + 1.01^2\ 2.33 \times 0.5/5.14) = 5.78\ \text{Gy}_{[5.14]}.$$

There is a deficit of 95.76-5.78 = 89.98 $\text{Gy}_{[5.14]}$.

This can be compensated for in the next 12 fractions, now with RBE parameters at a higher value of LET, by solving

$$12\ d(2.48 + 1.03\ d^2/5.14) = 89.98,\ \text{where}\ d = 2.5\ \text{Gy}.$$

For worked example B. For the critical volume of normal tissue, the megavoltage photon α of 0.1 Gy^{-1}, the relevant parameter changes are given in table 11.2.

The intended BED is 38.97 $\text{Gy}_{[3]}$, using a dose per fraction of 1.33 Gy, and this represents tolerance using the fixed value of RBE given. The 150% dose error will cause the dose per fraction to rise to around 2 Gy.

Table 11.2. Calculated parameters for worked example B.

Baseline Parameters	Planned treatment (LET = 15 keV . μm^{-1})	By Error (LET = 45 keV . μm^{-1})
$\alpha_L = 0.1\ (\text{Gy}^{-1})$	$\alpha_U = 0.906$	$\alpha_U = 0.906$
$\beta_L = 0.033\ (\text{Gy}^{-1})$	$\alpha_H = 0.15$	$\alpha_H = 0.249$
	$\text{RBE}_{max} = 1.50$	$\text{RBE}_{tnax} = 2.49$
	$\beta_U = 0.04$	$\beta_U = 0.04$
	$\beta_H = 0.0334$	$\beta_H = 0.0343$
	$\text{RBE}_{min} = 1.007$	$\text{RBE}_{min} = 1.02$

The excess BED over the three incorrect fractions, if uncorrected, would be

$$3 \times 2\,(2.49 + 1.02^2.2/3) - 3 \times 1.33\,(1.50 + 1.01^2 \times 1.33/3) = 11.33\ \mathrm{Gy}_{[3]}.$$

If toxicity increases by 1% per unit BED, then an 11.3% increase in toxicity might occur if no correction is applied.

The above steps are informative, and useful for teaching and research purposes, but there are many pitfalls within the process, more so than in the much simpler corrections sometimes necessary for errors arising in photon based therapy.

11.3.5 The potential impact of erroneous fractions on tumour control

A further study based on problem A1, but now assuming that the same half of the volume of a tumour has been under-dosed on one or more sequential fractions. This has been simulated in greater detail with estimation of tumour control probabilities (TCP). Plots of the numbers of erroneous treatment against the expected TCP are shown in figure 11.3. Here several assumptions of different tumour RBE values have been made in addition to the RBE predicted by the model described in this report (purple data points). Detailed inspection will show that the dose of 70 Gy in 15 fractions is very forgiving for treatment errors in the case of photons (the green data points). This is because the tumour can be cured with doses of around 40–45 Gy in 15 fractions; but the excess dose guarantees a high cure rate even when dose delivery is imprecise for up to ten fractions. When the tumour RBE of 2.5 (as stated in the problem) is itself incorrect, as shown by the red data points, the physical dose has been excessively reduced and even if all treatments are accurately delivered the TCP

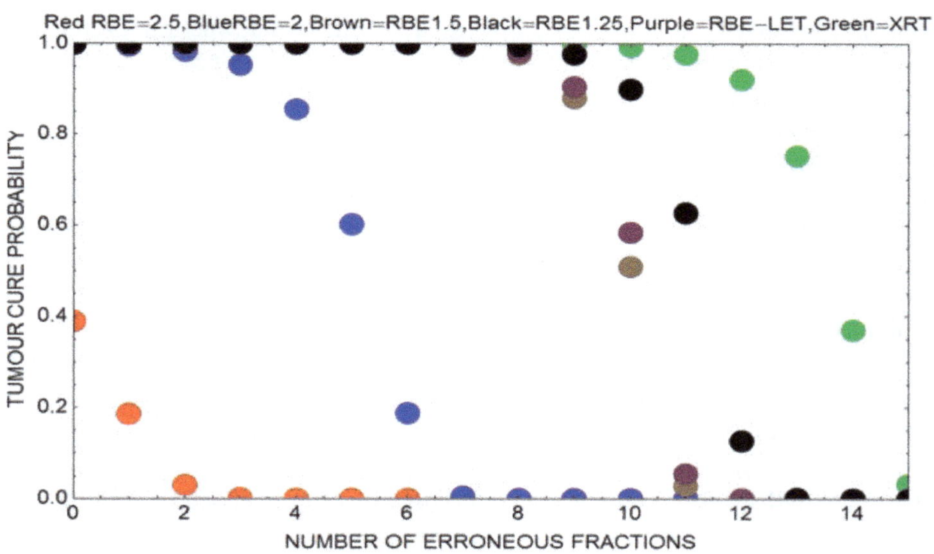

Figure 11.3. Plots of tumour cure probability against number of erroneous fractions where the same half of a tumour is under-dosed by 50%. The TCP is calculated using standard Poisson statistics with the same radiobiological assumptions as in example A1, with altered radiobiological parameters as in table 11.1, with 10^9 clonogens per tumour. No repopulation factor has been used since the overall time is constant and less than the approximately 28 day onset time of significant repopulation in a squamous cell carcinoma.

is reduced to around 0.4 (or a control rate of 40%) with a sharp fall-off in TCP if only one or two erroneous treatments are made. For more than two erroneous treatments the TCP is effectively zero. For lower assumed values of tumour RBE, the prescribed dose is not so reduced, allowing increasing TCP and also improving the numbers of erroneous treatments than can be made before causing a significant reduction in TCP.

In reality higher doses do oppose radio-resistance regardless of cause, whether intrinsic (e.g. genetic) or acquired (e.g. hypoxia). This example is sufficient to demonstrate that higher doses are protective against at least some forms of error in dose misplacement during particle therapy. Such higher doses can only be permitted if normal tissue tolerances of smaller volumes of critical organs at risk are respected, e.g. if the PTV includes normal tissue whose functional contribution can be lost, but without appreciable detriment in quality of life.

There are other situations in oncology, for example the treatment of radio-sensitive childhood tumours, where dose escalation to very high levels may not be acceptable since normal tissue tolerances of nearby structures have to be respected, e.g. the eye or spinal cord. In the context of more radio-resistant tumours, for example, a craniopharyngioma where optic nerve and optic chiasmal structures may extend into the CTV/PTV, dose escalation, if used at all, may have to be limited to a part of the tumour. In such situations, there is a risk of tumour under-dosage producing significant reduction in tumour control, which justifies increasing use of image guidance, especially if non-ionising radiation impact on patient throughput.

Just as in airports, in radiotherapy departments delays occur due to a variety of reasons, which are often connected with patient safety. Sometimes, interruptions during a treatment fraction occur, due to accelerator malfunction or the beam delivery process during therapy, although this is infrequent. In practice, a failed QA check before treatment has commenced leads to a delay in beginning or resuming treatments. Delays are sometimes simulated as having a Gaussian distribution but they are naturally bounded at the lower limit (because of more serious causes of failure, which need much longer times to be corrected), so will tend to have a log-normal distribution. Interruption of a treatment fraction requires a much more complicated approach, as shown above. Only if the treatment interruption is less than 0.5 hour should the remainder of the dose be applied without specific radiobiological correction. This is because DNA is repaired rapidly (in the CNS the fast component repair has a half time of 12 min and the slow component just over two hours (see chapter 2). The correction, if adopted comprehensively, will require input from physicists, clinical radiobiologists with an interest in modelling, and the responsible clinician.

The best that could be aimed for at the present time would be a 24 h break in treatment, although longer intervals may be anticipated in some instances. There are convenient ways to overcome longer delays after erroneous treatments have been delivered, if these occur early in a treatment course of many fractions, since the usual fractionation can be continued a day later, leaving the correction of the erroneous fraction (or fractions) to a later time, towards the end of a treatment course. This would allow the ‘correcting team’ a longer time to consider the problem; this may avoid the situation where a rapid but incorrect decision is taken after an erroneous fraction has been partly or totally given. In situations where voxel by voxel compensations are attempted this would require automated computing, with input of the relevant radiobiological parameters and many re-iterations to obtain a reasonable compromised solution.

Specific training in the clinical radiobiology area is required for several disciplines. Clinicians should be made more aware of the LET–dose–RBE issues and be encouraged to practice simple calculations with their physics and radiobiology colleagues.

This problem should be discussed in depth by the relevant bodies that govern the use of radiotherapy in each country. At the present time, full correction may take a physicist 4–6 h and the clinician (1–2 h) to discuss; if the problem requires several iterations these times will be far longer, taking perhaps a whole day or more.

Should a treatment error not be corrected until after the intended overall treatment time has elapsed, then repopulation related calculations may also be necessary, as described in chapter 10. This adds a further level of complexity where additional compromises will be necessary because the correct solution to preserve the same tumour control will not be the same as for normal tissue complications.

11.4 General discussion

The causes of failure in achieving local tumour control with radiotherapy are numerous, including intrinsic and acquired radio-resistance, patient related factors and dose delivery errors. At the present time, much attention is being given to quality assurance measures regarding dose and the accuracy of its delivery for each treatment. Further measures that minimise potential errors, such as adaptive radiotherapy (ADR) are by now in routine use in some treatment centres (Schwarz *et al* 2012). In ADR, the radiation fields are checked (using one or more of a variety of imaging procedures) at either frequent exposures, or even at each exposure and the treatment field directions/sizes modified to take into account changes in the patient, and tumour/normal tissue position and dimensions.

The Bragg peak position is dependent on the initial energy of the particle and its rate of energy loss with increasing tissue depth in the body; these parameters normally include some degree of variation, resulting in variations in the dose range. So, dose deposition consequently can be uncertain, especially if there are extremely non-homogenous tissue electron density variations in regions containing marked changes in chemical composition. CPT (charged particle therapy) is consequently more sensitive in terms of changes in dose placement during therapy. Even loss or gain in patient weight during treatment, or daily set up errors, could influence particle range, causing simultaneous tumour under-dosage and an increase in normal tissue dose compared to what was intended. Use of ultrasonic measurements of distances between skin and bone on a daily basis prior to entering the treatment room would help to reduce, for example, errors due to loss/gain in subcutaneous fat.

It is important to consider the potential effects of serial under-dosage, even if this only affects part of a tumour volume and to study what the influence of this will be on tumour control following conventional therapy and CPT.

The results show that increasing numbers of repeated under-dosage of the same section of a tumour can have deleterious consequences for tumour control probabilities.

Practical calculations should include the possibility of lower RBEs due to lower LET values in under-dosed regions using the passive scattering technique. This would need to be done on a voxel by voxel basis and is beyond the scope of the present study. A further important issue is the finding that for scanned beams that allow intensity modulated proton therapy there can be higher LET values (and so RBE values) just outside the PTV, where dose is reduced (Grassberger *et al* 2011). Both low dose and higher LET will lead to a higher RBE.

Adequate quality assurance and vigilance is required during radiotherapy to ensure that systematic under-dosage does not occur. This could occur due to a variety of reasons ranging from dose/particle numbers and energy/ranges as well as errors due to scattering or scanning of the beam. Patient set up errors, tissue depth and density variations, tumour shrinkage and other effects such as tumour motion with respiration may all contribute.

These issues impinge on medical ethics: how much of the uncertainties in the physical and biological should be discussed with individual patients? Should full informed consent include these in comparison with photon based techniques? Should more phase III randomised studies be done because of the influence of these uncertainties? Will patients who develop disappointing results take legal action on the basis that they were not fully informed as a natural human right, encoded in European Law?

The results presented are dependent on the validity of the linear-quadratic model and the various assumptions made. It will be important to collect more data on errors or interruptions during particle therapy, the net effect on LET distortions and estimated RBE changes, their correction details and to correlate this information with tumour control outcomes and also toxicity since overshooting of Bragg peaks may result in excess normal tissue dosage. A pan-European data base could be developed for this purpose.

The computing support required for LET based corrections of errors is considerable, with use of Monte Carlo systems which incorporate radiation transport data and which can interact with treatment planning information such as differences in tissue composition. In the short term, the only means of achieving this would be to use certain reference laboratories that are capable of achieving this and which could work with the software divisions of treatment planning companies. There should be at least one location in Europe and the USA, which could be an advisory centre for treatment interruptions of any kind. The use of cloud computing will undoubtedly help in this area, although anonymity of patient data is a legal requirement, which also makes patient data transfer difficult in some parts of the world. Written permission obtained from patients before treatment is commenced would probably be advantageous in this respect.

11.5 Conclusions

In order to achieve the best clinical outcomes following charged particle therapy, it is important to

- eliminate errors of dose delivery where possible by optimised treatment planning, good beam quality assurance, use of image guidance and, where possible, external confirmation of Bragg peak positions/dose distributions during particle therapy;
- correct significant errors by using rational strategies based on LET, RBE and BED to restore the intended tumour control following tumour under-dosage, or to reduce normal tissue effects if they have been unintentionally over-dosed.

The radiation oncologist/clinical radiobiologist and physicist must liaise closely in order to achieve these aims. The calculation of a corrective dose should involve BED equation (11.1) after detailed analysis of the original and erroneous treatment dose and LET distributions. A new system which links LET to RBE for clinical applications using any ion species (including protons) has been developed. At the present time there is no routine application of LET mapping in commercial software systems for routine proton treatment planning. LET maps leading to RBE maps and equivalent dose estimations, all in 3D, are now urgently required. This would allow rational compensatory calculations to be used. Such a development would, even in the absence of delivery errors, improve the clinical outcomes of routine particle therapy, since the clinician would be alerted about risks associated with higher and lower LET areas. The methods presented in this report provide a relatively simple framework for doing this, although it involves many more steps and assumptions than is the case for similar protocol deviations in megavoltage photon therapy.

Ideally, a single reference centre (or virtual centre) in Europe should be able to handle difficult errors efficiently, until necessary training has been undertaken. More advanced training programmes for clinicians and medical physicists are required in this area; their present training does not include adequate time or development of competencies in this particular area.

References

Grassberger C, Trofimov A, Lomax A and Paganetti H 2011 Variations in linear energy transfer within clinical proton therapy fields and the potential for biological treatment planning *Int. J. Radiat. Oncol. Biol. Phys.* **80** 1559–66

Jones B, Carabe-Fernandez A and Dale R G 2006 Calculation of high-LET radiotherapy dose required for compensation of overall treatment time extensions *Brit. J. Radiol.* **79** 254–7

Jones B and Dale R G 2008 Radiobiological compensation of treatment errors in radiotherapy *Brit. J. Radiol.* **81** 323–6

Schwartz D L *et al* 2012 Adaptive radiotherapy for head-and-neck cancer: initial clinical outcomes from a prospective trial *Int. J. Radiat. Oncol. Biol. Phys.* **83** 986–3

Weyrather W K, Ritter S, Scholz M and Kraft G 1999 RBE for carbon track-segment irradiation in cell lines of differing repair capacity *Int. J. Radiat. Biol.* **75** 1357–64

Chapter 12

Additional considerations and conclusions

12.1 Introduction

It is clear that there is ample scope to improve radiotherapeutics by using proton and ion beams (PIBT), but in what situations may the impact be greatest, and what can be done to reduce some of the existing disadvantages?

In radical treatments, PIBT will reduce collateral radiation effects in tissue distant from the cancer, which will be most important for long term survivors; those will low metastatic rates but achievable high local control rates, and where late vascular and fibrotic complications, as well as the bonus of reduced malignant induction probabilities, will all contribute to quality of life expectations.

A balance has to be struck between these benefits and the detriments that could follow the relative biological effect (RBE) uncertainties, the latter needing to be minimised for optimal normal tissue outcomes. The first priority in medicine is to do no harm (*primum non nocere*). Consequently adjustment of dose to be within normal tissue limits, where retention of that issue function is important, has to be the first priority. In some clinical situations, further radiation dose escalation may not necessarily produce greater cure rates, because of the influence of dose modifying and other adjuvant therapies. It is especially important that any radiation technique should not allow under-dosage (or 'geographical miss') of the cancer, and that the allocated RBE is not as unfavourable as to result in a lower than intended tumour dose. There may be special instances where a simpler PIBT technique will produce better dose distributions than any photon technique, as in cases where metallic bearing prostheses exist (Jones 2006, 2008), or to avoid the heart, in re-treatment situations, in palliative radiotherapy and other situations where congenital anatomical changes, or malformations, complicate treatment planning.

The role of PIBT in palliative and/or re-treatment situations (the latter may be given with radical or palliative intent), is often neglected due to assumed costs and the aim to obtain prolonged survival using radical treatments. Some new proton centres appear to report that up to 30% of patients belong in this category. After all,

doi:10.1088/978-0-7503-1338-4ch12

why should a patient receive a suboptimal dose distribution using photons when a better PIBT alternative is available and with much lower expected morbidity, especially if the treatment can be given efficiently in a limited number of fractions? Consider some practical examples: treatment of (a) para-aortic nodes, or (b) bone metastases in the cervical spine. In (a) the photon beam exit doses are considerable whatever photon-based technique is used, and the treatment volumes can be large so that standard fractionation may be required (1.8–2 Gy day^{-1}) over 4–5 weeks typically, although more focused techniques can allow hypofractionation but with a considerable 'dose bath' with resultant acute effects. In (b) the exit doses to the throat and upper oesophagus can cause prolonged mucositis, with pain on swallowing, which would not occur with carefully planned PIBT. Although some modern photon techniques can treat the metastatic tumour alone, with good initial response, there is often a need to re-treat in, or close to, the same volume, or there may be several nearby areas that require treatment in an '*en-bloc*' fashion.

Obtaining the most correct RBE for any particular situations remains a difficult hurdle, although guidance provided by careful modelling studies has probably improved this situation. In the future there may be further predictive assays which feed into RBE values beyond what is now possible by taking assumed α/β ratios (as used in standard clinical radiobiological modelling), or likely values of α and β, especially since gene expression patterns and DNA methylation are markedly different after protons in comparison with conventional therapy x-rays (Girdhani *et al* 2013, Chaudhary *et al* 2014, 2015). More clustered DNA damage in Bragg peak regions influence cell killing and ultimately the RBE dose ratio. It is possible that indications for proton beam therapy in certain tumour types may alter on this basis. It is also intriguing that the oxygen enhancement ratio may also reduce in the Bragg peak regions and could be close to those for fast neutrons, since neutrons cause ionisation mainly by releasing recoil protons. This again makes a strong case for the high linear energy transfer (LET) regions to be maintained in the gross tumour volume, and not in normal tissue regions.

The remainder of this final chapter considers a miscellany of different but novel approaches:

12.2 Dose escalation where circumstances permit

The possibility of increasing the prescribed tumour dose when it is judged that the degree of important normal tissue sparing (outside the planning target volume (PTV)) is good, but where dysfunction of the normal tissue within the PTV will not significantly influence subsequent quality of life. Here it is assumed that the main organ at risk outside the PTV is taken to close to its tolerance in order to increase tumour dose as much as possible. This strategy links with the optimum dose per fraction concept presented in chapter 3.

Tables 12.1 and 12.2 present estimated proton dose increases that may be possible for a radioresistant and a radiosensitive tumour, respectively, in the case of a 1.1 RBE and the variable RBE model, where X is the degree of sparing (only $X = 1$ (no sparing) and $X = 0.8$, or 80% sparing are considered). Four different fractionation

Table 12.1. 'Radioresistant' example. An asterix is shown where tissue tolerance has been exceeded.

	All RBE = 1.1 (the present condition)			'dose escalated' variable RBE			
Dose (Gy) per #, N, and X	NTBED (EQD-2) in PTV	NTBED (EQD-2) OTV	TUMBED (EQD-2) in PTV	NTBED (EQD2) in PTV	NTBED (EQD-2) OTV	TUMBED (EQD-2) In PTV	Modified dose/#
2, 25, 1	108.61* (54.30)	N/A	63.90 (48.35)	100.0 (50.0)	N/A	58.92 (44.59)	1.71
2, 25, 0.8	N/A	80.38 (40.19)	63.90 (48.35)	N/A	96.99 (48.50)	77.09 (58.34)	2.09
2.75, 15, 1	102.18* (51.09)	N/A	58.85 (44.53)	97.97 (48.98)	N/A	56.61 (42.84)	2.43
2.75, 15, 0.8	N/A	74.36 (37.18)	58.85 (44.53)	N/A	94.58 (47.29)	73.89 (55.92)	2.96
4.8, 6, 1	95.34 (47.67)	N/A	50.47 (38.19)	97.92 (48.96)	N/A	51.66 (39.09)	4.44
4.8, 6, 0.8	N/A	67.28 (33.64)	50.47 (38.19)	N/A	93.55 (46.77)	67.95 (51.42)	5.39
9, 2, 1	90.32 (45.16)	N/A	42.07 (31.83)	99.0 (49.5)	N/A	45.57 (34.48)	8.63
9, 2, 0.8	N/A	61.7 (30.86)	42.07 (31.83)	N/A	93.46 (46.73)	61.05 (46.20)	10.43

N/A = not applicable.

Table 12.2. Radiosensitive example.

	All RBE = 1.1 (the present condition)			DOSE ESCALATED VARIABLE RBE			
Dose (Gy) per #, *N*, and *X*	NTBED (EQD-2) in PTV	NTBED (EQD-2) OTV	TUMBED (EQD-2) in PTV	NTBED (EQD2) in PTV	NTBED (EQD-2) OTV	TUMBED (EQD-2) in PTV	Modified dose/#
2, 25, 1	108.61* (54.30)	N/A	44.43 (41.41)	100.0 (50.0)	N/A	40.90 (38.13)	1.71
2, 25, 0.8	N/A	80.38 (40.19)	44.43 (41.41)	N/A	96.99 (48.49)	53.59 (49.96)	2.09
2.75, 15, 1	102.18* (51.09)	N/A	40.72 (37.96)	97.97 (48.98)	N/A	39.24 (36.58)	2.43
2.75, 15, 0.8	N/A	74.36 (37.18)	40.72 (37.96)	N/A	94.58 (47.29)	50.40 (46.98)	2.96
4.8, 6, 1	95.34 (47.67)	N/A	32.79 (30.57)	97.92 (48.96)	N/A	33.46 (31.19)	4.44
4.8, 6, 0.8	N/A	67.28 (33.64)	32.79 (30.57)	N/A	93.55 (46.78)	42.35 (39.48)	5.39
9, 2, 1	90.32 (45.16)	N/A	23.93 (22.30)	99.0 (49.5)	N/A	25.53 (23.08)	8.63
9, 2, 0.8	N/A	61.7 (30.86)	23.93 (22.30)	N/A	93.46 (46.73)	32.36 (30.17)	10.43

N/A = not applicable.

schedules are used. It can be seen that the fixed 1.1 RBE can result in over-dosage when small doses per fraction are used, but the normal tissue (NT) biological effect dose (BED) is kept to within the tolerance of the critical NT outside the PTV with the variable model and which permits the use of the increased doses per fraction to the tumour target as shown. The modelling used here is the same as in chapters 9 and 10, but with assumed values of $\alpha = 0.55$ Gy^{-1} and $\beta = 0.02$ Gy^{-2} ($\alpha/\beta = 27.5$ Gy) for the radiosensitive category, with $\alpha = 0.28$ Gy^{-1} and $\beta = 0.045$ Gy^{-2} ($\alpha/\beta = 6.2$ Gy) for the radioresistant category; in each case the normal tissue had assumed $\alpha = 0.06$ Gy^{-1} and $\beta = 0.02$ Gy^{-2} ($\alpha/\beta = 2$ Gy).

12.3 Simultaneous 'sensitisation' effects by new therapies

The considerable advances made with descriptive molecular biology and its applications in cancer research provide new opportunities for molecular modification of radiation responses. These may be useful within the timeframe of a radiotherapy treatment courses, provided that any positive tumour effect exceeds any deleterious effects on key normal tissues. Quantification of such effects remains controversial, and in the context of using such approaches with charged particle beams, there are additional complexities due to the RBE issues. The sensitizer enhancement ratio (SER) is defined as the radiation-only dose required for the same biological effect (without the sensitising agent) divided by the radiation dose for the same effect when the sensitiser is present. It is analogous to the RBE definition, which utilises horizontal shifts of the cell survival curves.

Following discussions with Professor R G Dale, in order to find the vertical shifts in the cell survival curves due to a sensitising drug, it is suggested that a BED approach is taken in order to accommodate the RBE and the SER. The theoretical cell survival curves in figure 12.1 refer to the use of a drug (D) to sensitise (1) standard megavoltage photon therapy and (2) a high LET therapy, e.g. carbon ions.

It is assumed that the drug sensitises the low LET α_L and β_L values by factors of a and b, respectively, but the drug changes the high LET radiosensitivities by factors of A and B. For example the α_L becomes $a\,\alpha_L$, which can be renamed as ${}_s\alpha_L$, (the sensitised α at low LET), which will be the parameter measured by the cell biologist. The same terminology applies for β, and the process is extended to the high LET state, as shown in table 12.3, which gives the relevant equations for sensitisation, including the radiosensitivity changes and their BED equivalents for the purposes of future analysis.

In table 12.4, the (advantageous) incremental ratios of the BEDs due to sensitisation are shown for different radiation doses and this approach would be reasonable for the analysis of clinical results. Figure 12.2 shows the effective sensitiser enhancement ratios achieved with the combined effects of the drug and the two classes of radiation as a function of dose. The form of these curves depends on the degree to which α and β are sensitised. In general it should be remembered that sensitisation of α (the dominant effect in high LET radiations), has greater biological effects at low dose per fraction, but β sensitisation confers more effects with increased dose per fraction.

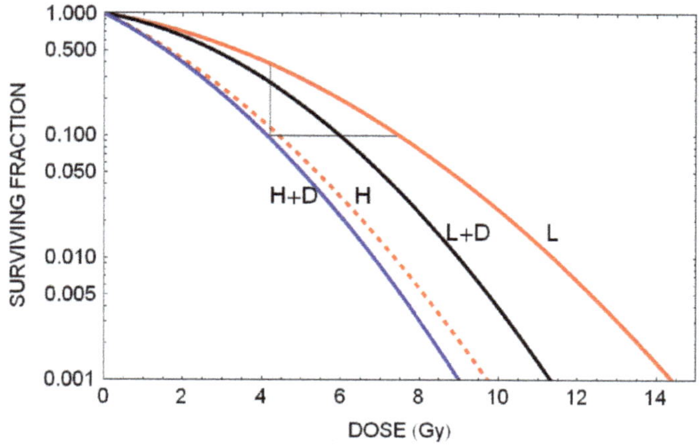

Figure 12.1. Cell survival curves for assumed parameters of: $\alpha_L = 0.12$, $\beta_L = 0.025$; $a = 1.1$, $b = 1.3$; $A = 1.05$, $B = 1.2$; $RBE_{max} = 3$, $RBE_{min} = 1.2$. L refers to the low LET reference radiation, H to a higher LET using carbon ions, and D refers to the sensitising drug. The horizontal line at a surviving fraction of 0.1 can be used to estimate RBE and the sensitiser enhancement ratios. In contrast, the incremental BED ratio method described in the text reflects the vertical changes between the four curves, as represented by the vertical line at a dose of 4.2 Gy.

Table 12.3. Lists of surviving fraction and BED equations for each category shown in figure 12.1, where the low LET α/β ratio is now represented by k. The composite (upper) and measured (lower) radiosensitivities in each category are shown in column 2, with BED representations in column 3.

Radiation and drug category	$-\log_e$ (surviving fraction)	BED
Low LET only	$N(\alpha_L d + \beta_L d^2)$	$D(1 + d/k)$
Low LET + D	$N(\alpha_L.ad + \beta_L b_2 d^2) = N({}_s\alpha_L d + {}_s\beta_L b^2 d^2)$	$D(a + b^2 d/k)$
High LET only	$N(\alpha_L R_{max} d + \beta_L R_{min}{}^2 d^2) = N(\alpha_H d + \beta_H d^2)$	$D(R_{max} + R_{min}{}^2 d/k)$
High LET + D	$N(\alpha_L R_{max} A d + \beta_L R_{min}{}^2 B^2 d^2) = N({}_s\alpha_H d + {}_s\beta_H d^2)$	$D(R_{max} A + R_{min}{}^2 B^2 d/k)$

Table 12.4. BED increment ratios for four treatment conditions.

BED ratios for:	$d = 2$ Gy	$d = 4$ Gy	$d = 6$ Gy
1. High LET/low LET	2.54	2.29	2.13
2. Low LET + SD/low LET alone	1.27	1.37	1.43
3. High LET + SD/high LET alone	1.12	1.16	1.20
4. Combined high LET + drug sensitised/low LET alone	2.83	2.66	2.55

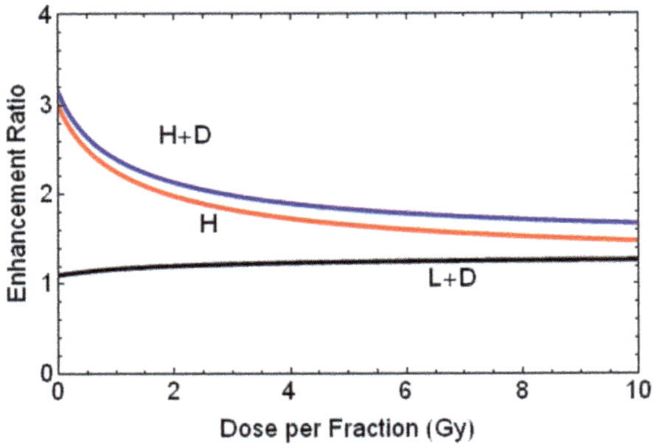

Figure 12.2. Plot of enhancement ratios changing with physical dose per fraction for three conditions of enhancement where L + D is drug with low LET radiation, H is high LET alone and H + D is high LET radiation plus sensitising drug.

Table 12.5. How experimental data can be used to determine sensitisation parameters R_{max}, R_{min}, a, b, A and B.

Experimental data obtained	Output
Low and high LET radiation (both with no SD) cell survival curves experiments provide α_H and α_L parameters	R_{max} (= α_H/α_L) and R$_{min}$ (= $\sqrt{\beta_H}/\sqrt{\beta_L}$)
Low LET with SD provide $_s\alpha_L$ and sβ_L parameters	a (= $_s\alpha_L/\ \alpha_L$) and b (=$\sqrt{_s\beta_L}/\sqrt{\beta_L}$)
High LET with SD provide $_s\alpha_H$ and $_s\beta_H$ parameters	A (= $_s\alpha$H/ αH) and B (=$\sqrt{_s\beta_H}/\sqrt{\beta_H}$)

A protocol for the analysis of low and high LET radiation with sensitising drugs is provided in table 12.5 using the same four cell survival curves, as described for figure 12.1. Such approaches, which may show striking advantages with respect to tumour cell killing must, of course, be repeated using acute and late reacting normal tissue assays (not only acute reacting tissues, as was the case with so many RBE studies in the past).

12.4 Sensitivity analysis of the energy efficiency model

For the model presented in chapters 8 and 9, the dependency of the LET$_U$ position and the degree of saturation of α and β with increasing LET will all influence the RBE estimates, as shown in figure 12.3(a)–(c) for reasonable shifts in the assumed values used in this book, which are the central values given. The changes, although small, can be of clinical significance, more so at low dose for the α-related shifts and at high dose for the β-related shifts and at lower doses for the LET$_U$ position. It is important that model input parameters are updated in accordance with new

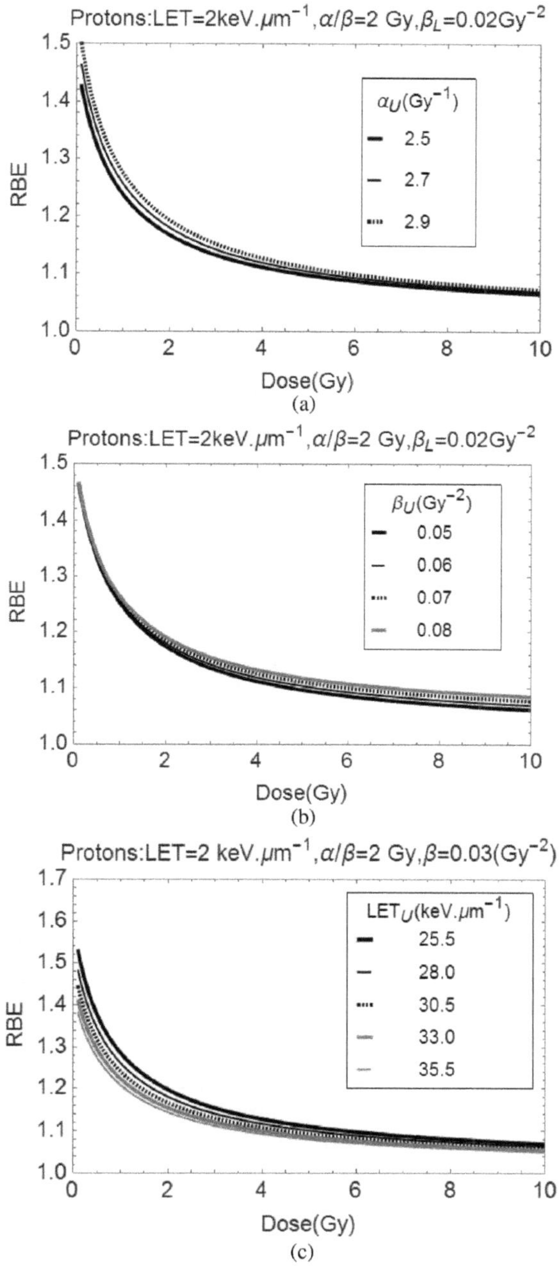

Figure 12.3. (a)–(c) Plots of RBE with dose per fraction showing sensitivity to changes of assumed α_U, β_U and LET_U, the maximum efficiency point LET_U, for $\alpha/\beta = 2$ Gy where LET is 2 KeV mm^{-1}.

information. Further detailed research would be required to refine the precision of these parameters, as indicated below.

It is clear that small refinements may be possible and can only be achieved by experimentation. To find the true LET–RBE turnover points it is also necessary to determine both α and β changes with LET, so that the entire model may be validated and compared with other models. This might involve a very large international collaboration, but preferably conducted in one laboratory with the capacity to study a range of ionic beams at different energies and with delivery systems where the influence of inter track distances on biological effects can be studied, where the tracks are in the same primary direction and in with other geometrical combinations (e.g. orthogonal or opposed tracks) by varying the irradiation conditions.

12.5 What could be achieved in a single international laboratory dedicated to high LET radiobiology

The proposed Bio-LEIR project at CERN (Dosanjh *et al* 2013) is entirely capable of addressing the above issues. Just as in the case of the CERN LHC experiments, which confirmed the presence of all previously known subatomic particles during the first six weeks of energy ramp-up, and to a much greater level of parameter precision then achieved previously, bio-experiments could also benefit from a carefully planned experimental programme to confirm previous phenomena, but also obtain large data sets with the aim of finding robust and more accurate parameters for predictive modelling purposes. Some examples of what might be achieved are discussed below.

12.5.1 Simulated experiments

Simulations have been performed of future experiments designed to determine the LET–RBE turnover point positions at $LET_{U,}$ (and for the underlying LET-α and LET-β relationships), for protons and ions with different Z numbers. The methods are as follows.

1. Use random sampling computer techniques with typical coefficients of variation for cell survival studies to provide α and β values at the control and reference LET values. The number of different cell survival experiments can be varied.
2. Then vary the number of experiments for accurate estimation of the two linear gradients of α and β (separately) with LET for lines before and beyond LET_U, in order to determine their intersection position at LET_U with a high degree of accuracy for each Z number. The α and β parameters obtained may be combined to give an RBE value for each given dose.
3. Confirm if LET_U remains constant at three to four different dose levels, and for different cell lines, for each different ion (Z value).
4. Confirm and strengthen the precision of the mathematical relationship between Z and LET_U, with Monte Carlo simulations of these.
5. Estimate the increased precision of the gradient of α and β with LET and determine their maximum values α_U and β_U at LET_U for each Z number, and for different cell types.
6. Repeat the above in several typical beam depth dose positions and for different incident energies.

7. Apply statistical goodness of fit tests to find the best of the existing LET–RBE models, or which models fit the data best in what situation, e.g. low/intermediate/high dose.
8. Estimate to what extent the better prediction of RBE would improve results of potential clinical applications.

Figure 12.4 shows one example of a simulation.

Four repeated simulations, using seeded random sampling, provide LET_U values of 29.6, 32.06, 33.73 and 27.64, yielding a mean of 30.75 ± 1.34 (SEM) when there are 20 LET-RBE data points. The output of one such simulation is given in figure 12.4. The simulations were set up with an assumed real LET_U value of 30.5. This is a reasonably good result for medical purposes, but if the numbers of LET-RBE data points are reduced to 16 the estimate falls to 29.96 ± 1.24, and if further reduced to 12 the estimate is 31.5 ± 1.92. The last two examples were simulated with fewer repeat experiments leading up to the RBE determination (three repeats rather than four for each irradiation and six points rather than eight on the reference radiation survival curve). Further reductions in the number of repeat cellular experiments required to produce each RBE data point show even more unreliable results. It follows that very large experiments are necessary to produce the most reliable data, although it is important at this stage to assess the clinical significance

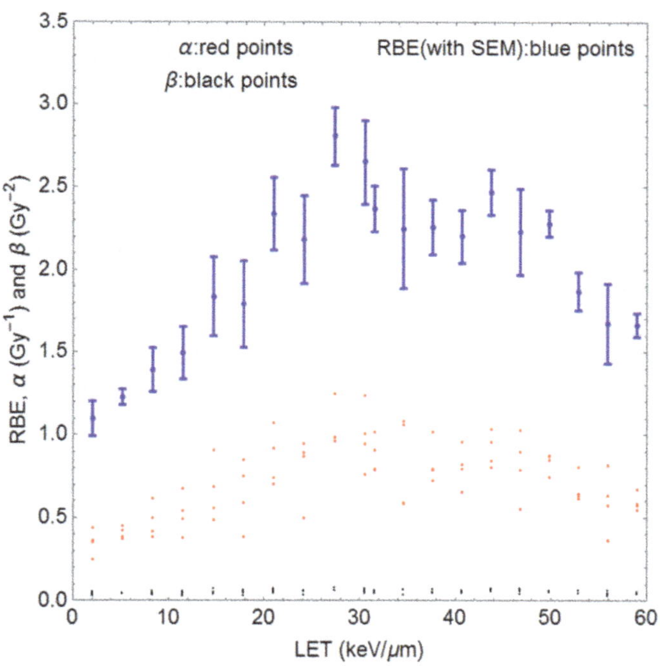

Figure 12.4. Example of a single *Mathematica* (Wolfram, USA) simulation of a LET_U determination experiment for protons, using variations in cellular radiosensitivities for cell survival assays. Here 20 RBE data points are used. The LET_U value is obtained by obtaining the intersection point of the two upward and downward data linear regression fits. The expected LET_U is assumed to be 30.5 keV μm^{-1}, but the fitted value for this simulation is 27.64 (owing to a coefficient of variance of 20% for all radiosensitivity measurements).

of an error in estimating LET_U. From geometrical considerations, an experimental LET_U result of say 28.8 or 31.8 keV μm^{-1} (instead of the assumed correct 30.5) would lead to an approximately 5% error in RBE estimation. However, by feeding this result into clinical iso-effect equations using the BED concept, in which the RBE is represented by its maximum and minimum limits at any specified LET,

$$BED = nd_H\left(RBE_{max} + \frac{RBE_{min}\, d_H}{(\alpha/\beta)_L}\right),$$

where n is the number of doses given and the α/β ratio is that of the reference (or control) radiation modality, the change in the recommended dose for iso-effective neurological damage reaches a 5% difference when the estimated LET_U value falls down to between 25 and 26 keV μm^{-1} (around a 17% change in LET_U), when the local LET is as high as only 9 keV μm^{-1}, a value which can appear on proton therapy 3D LET maps. So, these estimates suggest that the magnitude of the error is not fully propagated in a practical clinical situation, where tissue is treated from multiple directions using pencil beams and where the LET values are not close to their maximum possible values. However, small deviations of dose beyond the already allowed ICRU dose variation of −5% to + 7% across a clinical target volume (consisting of the cancer plus surrounding normal tissue at high risk of harbouring infiltrating tumour cells) might also contribute to tissue complications. In general terms, this would be at a rate of around 1%–2% per additional Gray, and so an additional 5% difference could mean up to 10% or more increased tissue complications and also possibly a reduction of tumour control if the error propagates as a dose reduction to the tumour, at around a 1% per Gray. These differences may seem small but could be of important clinical significance.

However, the situation is further complicated by the present clinical practice of allocating a constant RBE value of 1.1 for protons in all tissues regardless of dose and LET. Then, for a LET of 9 keV μm^{-1}, the discrepancies between the calculated and normally given dose (normalised to 100%) are: 80.21% for use of the correct LET_U, and 76.92% for an incorrect LET_U of 27 keV μm^{-1}. So, a correct allocation of LET_U would make a further dose change of around 4.2% over and above the large difference due to a fixed versus flexible RBE allocation. This shows the considerable potential importance of using a variable LET-based RBE system with an accurate LET_U value. To change the present assumption of a constant proton RBE regardless of LET, dose and cellular system, would require a large series of careful experiments.

The same considerations would apply for heavier ions where the spread-out Bragg peak values of, say, C ions are normally at a LET that is around 20%–30% of the LET_U value, whereas the above proton example had a LET of up to 9 keV μm^{-1}, i.e. around 30% of LET_U. Much more detailed work is required to pursue these aspects further and for each ion species.

12.5.2 Uniqueness of LET_U for each ion species

The uniqueness of LET_U for each ion species is now considered further. This hypothesis can be addressed only to a certain extent from existing data (see Jones

2015), owing to there being relatively few reliable data points available (for heavier ions the turnover points are not so well defined due to data and range limitations, as can be visualised in Sorensen *et al* 2011). Only by combining results from past experiments using different cell types and at different doses can the LET_U position be studied at the present time, but then it is not possible to separately test the other hypotheses mentioned earlier.

Two such examples, for C ions and helium ions, are shown in table 12.6, where statistically significant differences are obtained. Rigorous experimental proof of unique LET_U turnover points for each ion species would oppose many existing statements made that a LET value of around 110 keV μm^{-1} represents the maximum obtainable RBE for all ion species, as stated in many textbooks, for example Hall and Gaccia (2011).

Further pragmatic experiments linked with theory could be performed, for example.

- Investigate the reduction in RBE with depth reported in France (F Megnin-Chanet's experiments reported by Calugaru *et al* (2011)) with a greater number of data points and for different cell types and doses, and compare to Monte Carlo simulations. This is likely to require an additional parameter beyond radiation fluence, KERMA, absorbed dose and LET, which is defined in one plane only; the possibility of quantifying the mean inter-track distance with depth needs to be considered, or by using a more volumetric interpretation of LET in x, y and z planes, due to increased Coulombic scattering with depth, especially in the case of protons, but less so with heavier ions. Indeed, this might be found to be a proton-only effect, but it is vital to determine this as treatment planning systems could incorporate such effects.
- The biological effects of nuclear fragmentation products can be separately studied.
- Establish comprehensive data sets for RBE_{max} and RBE_{min} values at different LETs for different ionic beams in panels of cells with known genetic characterisation.

Table 12.6. The locations of the combined C ion and helium data are significantly different ((Mann–Whitney $p = 0.028$, t-test $p < 0.0001$).

Ion and data source	Cell type	Estimated LET_U (keV μm^{-1}) (mean, standard error)
C ions, (Weyrather *et al* 1999) GSI, Darmstadt, Germany	CHO	145.81 ± 9.88
	V-79	159.05 ± 3.95
	Combined CHO + V-79 data	152.43 ± 4.29
Helium, (Barendsen 1968) Netherlands	Human T cells	124.24 ± 0.56

12.5.3 Priority in radiobiological experiments

The priority would be to establish the 'ground rules' for RBE determination but with greater precision and in a wider range of cellular systems. Later advanced experiments in collaboration with European Universities would inevitably follow, examining cellular signalling and damage response mechanisms at the molecular level, and their modification. It is important to point out that during the more basic RBE experiments, some cellular material can be stored for analysis at other centres of excellence with great expertise in molecular mechanisms. Thus the experiments could provide data for several purposes, as determined by an overall controlling committee structure. Other universities are already aiming to sensitise/modify radiation effects, or personalise radiation by attention to genetic and proteomic screening, etc. Also, the capacity to do long continuous low dose rate exposures for nuclear medicine and eventually the use of mixed field radiations (combinations of ions), their modification by hypoxia and other molecular approaches would be of great interest. Any such experiments should be based on the secure foundation of more fundamental knowledge.

Before radiobiology studies can commence, it would be essential to begin with beam characterisation work, sophisticated dosimetry and confirmation of dose in humanoid tissue equivalent 'phantom' structures, to confirm the numbers of particles, their ranges with depth and to translate the energy released and absorbed to the absorbed dose concept and determined to the quality standards demanded in medical practice and approved by registered medical physicists. The experience gained would be invaluable and could be used to establish a dosimetry reference laboratory for particle beams working closely with national standards laboratories and international bodies such as the ICRU and ICRP. An entire range of diagnostic and dosimetry equipment would be necessary for routine dose monitoring and would present the opportunity for advanced detector research and development, using direct beam monitoring, detectors inserted in tissue equivalent phantoms and remote detections of nuclear activation products.

After comprehensive studies of the beam ballistics for the various ions, radiobiological studies would follow using the same experimental beam geometry, which can be made to simulate the human body and the typical dimensions of treatments in various anatomical sites. The relationship between inter-track distances (related to fluence) and RBE needs investigation in detail (as suggested in chapter 1).

12.6 Conclusions

There is considerable scope for experimental radiobiological studies to refine knowledge of the RBE in protons and other heavier ions. These can be designed to improve the predictive modelling associated with particle physics applications in medicine by addressing fundamental issues of how particle physics interacts with bio-systems in a much more detailed way than previously attempted. The fulfilment of such a project would result in more controlled systems for safer and more effective proton and ion beam therapy. Until that can be achieved it will be necessary to make the best use of existing radiobiological RBE data sets.

References

Barendsen G W 1968 Responses of cultured cells, tumours and normal tissues to radiations of different linear energy transfer *Curr. Topics Radiat. Res.* Q **4** 293–356

Calugaru V, Nauraye C, Noël G, Giocanti N, Favaudon V and Mégnin-Chanet F 2011 Radiobiological characterization of two therapeutic proton beams with different initial energy spectra used at the Institut Curie Proton Therapy Center in Orsay *Int. J. Radiat. Oncol. Biol. Phys.* **81** 1136–43

Chaudhary P *et al* 2014 Relative biological effectiveness variation along monoenergetic and modulated Bragg peaks of a 62-MeV therapeutic proton beam: a preclinical assessment *Int. J. Radiat. Oncol. Biol. Phys.* **90** 27–35

Chaudhary P, Marshall T I, Currell F J, Kacperek A, Schettino G and Prise K M 2016 Variations in the Processing of DNA Double-Strand Breaks Along 60-MeV Therapeutic Proton Beams *Int. J. Radiat. Oncol. Biol. Phys.* **95** 86–94

Dosanjh M, Jones B and Myers S 2013 A possible biomedical facility at CERN *Brit. J. Radiol.* **86** 1025–9

Girdhani S, Sachs R and Hlatky L 2013 Biological effects of proton radiation: what we know and don't know *Radiat. Res.* **179** 257–72

Hall E J and Giaccia A 2011 *Radiobiology for the Radiologist* 7th edn (Philadelphia, PA: Lippincott Williams and Wilkins) chs 6 and 7 253–70

Jones B 2006 The case for particle therapy *Brit. J. Radiol.* **79** 24–31

Jones B 2008 Personal View: The potential advantages of charged particle radiotherapy using protons or light ions *Clin. Oncol.* **20** 555–63

Jones B 2015 A simpler energy transfer efficiency model to predict relative biological effect (RBE) for protons and heavier ions *Front. Oncol.* **5** 184 erratum 2016 *Front. Oncol.* **6** 32

Sørensen B S, Overgaard J and Bassler N 2011 *In vitro* RBE-LET dependence for multiple particle types *Acta Oncol.* **50** 757–62

Weyrather W K, Ritter S, Scholz M and Kraft G 1999 RBE for carbon track-segment irradiation in cell lines of differing repair capacity *Int. J. Radiat. Biol.* **75** 1357–64

www.ingramcontent.com/pod-product-compliance
Ingram Content Group UK Ltd.
Pitfield, Milton Keynes, MK11 3LW, UK
UKHW051855100926
8556IPUK00005B/638